Kao Gong Ji

Technology and Change in History

Series Editors

Adam Lucas
Steven A. Walton

VOLUME 17

The titles published in this series are listed at *brill.com/tch*

Kao Gong Ji

The World's Oldest Encyclopaedia of Technologies

Translated and commented by

Guan Zengjian
Konrad Herrmann

BRILL

LEIDEN | BOSTON

Revised and updated translation from the Chinese edition:《考工记: 翻译与评注》(ISBN: 978-7-313-12133-2) by Guan Zengjian, © Shanghai Jiao Tong University Press 2014. Published by Shanghai Jiao Tong University Press. All rights reserved.

This edition is jointly published with Shanghai Jiao Tong University Press.

The Library of Congress Cataloging-in-Publication Data is available online at http://catalog.loc.gov

Typeface for the Latin, Greek, and Cyrillic scripts: "Brill". See and download: brill.com/brill-typeface.

ISSN 1385-920X
ISBN 978-90-04-41690-1 (hardback)
ISBN 978-90-04-41694-9 (e-book)

This book is printed on acid-free paper and produced in a sustainable manner.

Contents

Abbreviations and Conventions

AC ancient Chinese
MC modern Chinese

For the transcription of Chinese names the Pinyin transcription of the P.R. China is used.

Illustrations

Tables

Preface
前言

在世界诸文明中，中华文明最为独特，这是因为，只有中华文明连绵不绝，可以一直向前追溯 5000 年。与之相应的是，有大量文字记录的文化被一代代传递了下来。在古代中国，尤其是在汉朝 (206 BCE–221 CE) 独尊儒术之后，统治者习惯于用文化来约束和引导社会精英的思维模式，以此统一人们的精神世界和道德准则。由此，对于中国古代社会来说，哲学和历史书籍具有至为重要的地位，这是不难想象的。与之形成对比的是，包含科技内容的书籍流传下来的则相对较少。

另一方面，战国时期 (475–221 BCE) 却是一个特例。在中国历史上，该时期以其科学技术的飞速发展而著称。这一点已经被近几十年的考古发现所证实。中国古代有一些颇负盛名的技术书籍，《考工记》就是其中之一。该书被保存于《周礼》一书的第六章，也就是其最后一章中。

《考工记》是迄今所知中国最古老的技术书籍。该书的绝大部分内容约产生于春秋 (770–476 BCE) 战国 (475–221 BCE) 时期，这意味着它已经有了 2500 多年的历史。由此，在世界范围内，它也是一部最古老的技术书籍。

另一方面，在春秋战国时期的一些哲学书籍中，也能发现一些科技内容。在墨子所著《墨经》中，这一现象尤其明显。《墨经》的形成时间，与《考工记》基本是一致的。无论如何，《考工记》是第一部纯粹讨论技术内容的书籍。

《考工记》以经验知识为基础制订技术规范，为兵车制作、青铜技术、陶瓷制品以及弓箭制造等等奠定了基础。该书首次对不同的青铜器依据其不同的功能要求，规定了其所对应的铜和锡的比例。《考工记》描述了融长度、容积和重量为一体的复合标准器的结构，500 年后，王莽新朝依据《考工记》的记述，制作了著名的复合标准器 “嘉量”，该嘉量原物现在珍藏于台湾台北的故宫博物院中，供人们观赏。《考工记》对复合标准器的记述，在世界范围内，是最早的。《考工记》的生物分类思想也不同寻常，它认为，人也是生物群体中的一种。《考工记》还展示了当时人们使用的各种不同类型的陶瓷器，这在历史上也是首次。《考工记》中记载的城市建造规则，在后来中国的首都北京和其他一些城市的建造中，仍然可以看到。这些规则的目的是要维护社会秩序，它们是依据传统社会的要求，同时考虑到贸易和交通的需要而制订出来的。

《考工记》对于它所描绘的所有产品，都给出了制作该种物品所需要的材料的品种和其大小、体积和重量的具体数据，并对制作过程做了具体说明。同时，该书还多处提到如何对产品进行测试。正因为这样，该书作为一部技术书籍，同时也为生产的标准化、计量和质量保证奠定了基础。公元前 221 年，秦始皇统一中国，随后颁布了大量法令，要求书同文、车同轨、统一度量衡。显然，在此之前，中国社会在标准化、计量和质量保证领域的这一传统，已经存在了很久。

然而，长期以来，西方的科学技术史研究本质上囿于欧洲文明，只是到了近几十年，关于中国的科学技术史，英国化学家和科学史家李约瑟(1900–1995) 及其在中国和西方同事们做了大量的开拓性研究工作，这才使情况有了变化。人们对欧洲之外的文明，特别是对中国文明的兴趣，有了明显的增加。李约瑟的研究旨在探讨各种民族科学的发展对世界科学的意义，探寻民族科学彼此之间的关系。

鉴于《考工记》对中国古代技术的发展具有极为突出的重要意义，我们决定将其翻译为现代汉语、英语和德语，以飨全世界对科学技术的历史感兴趣的人们。这也是《考工记》在世界范围内首次被翻译成英语和德语。在翻译过程中，我们借鉴了闻人军教授对《考工记》长期而又富有价值的研究工作。在此，我们谨对闻人军教授表示深切的谢意。我们衷心地向 Matthew Klopfstein 博士和 Marieke Krijnen 女士博士润色英文翻译道谢。

另一方面，要将《考工记》翻译为现代汉语、英语和德语，首先要对《考工记》本身有所研究。实际上，我们在自己研究的基础上，对《考工记》多处文本都提出了新的解释。例如，在车的制作一章，关于车轮和车毂的尺寸问题；在青铜技术一章，关于出土的青铜器的铜锡比例与《考工记》的记载是否一致的问题；以及在王城中祖庙的设计的问题，等等，我们都得到了新的认识。我们期望未来的考古发现能够进一步澄清在《考工记》翻译中现存的那些问题。

愿我们的工作能对促进中国古代科技成就在世界的传播有所助益。

关增建
Konrad Herrmann
2018 年于上海

Foreword

Among the civilisations of the world, China is so far unequaled, since it is the only civilisation that can look back on a continuous tradition of 5,000 years. It has, accordingly, acquired a rich body of written culture, much of which remains untranslated and therefore unknown in the West.

Ancient China's rulers sought to align the ideology of the elite with a set of unified moral and ethical principles informed by officially sanctioned philosophical and historical works. This was especially true following the victory of Confucianism during the Han dynasty (206 BCE–221 CE). It is therefore not surprising that the ancient Chinese works known to us today primarily concern history and philosophy, whereas relatively few works containing scientific and technological content have been conveyed to us.

The Zhanguo period (475–221 BCE) distinguishes itself by a lively development of science and technology that has been confirmed not only by the profound research of Joseph Needham and his many collaborators, but also by diverse archaeological findings over the last few decades. One of the most important technological books of ancient China is the "Kao Gong Ji" (*The Artificer's Record*), which was, perhaps tellingly, "hidden" as the sixth and last chapter in the work "Zhou Li" (*Rites of Zhou Dynasty*), but rediscovered and commented upon already during the Han dynasty. It has so far however not reached Western audiences.

The "Kao Gong Ji" is the oldest book on technology known in China. Excerpts with technological content appeared only later in philosophical, mathematical or historiographic works. Because its largest part probably originated in the era between the Chunqiu (770–476 BCE) and the Zhanguo period (475–221 BCE) about 2,500 years ago, "Kao Gong Ji" is also the oldest surviving technological work in the world. Written sources on science and technology that are roughly comparable include the works, from a later moment in time, of the Greek mathematician, physicist and engineer Archimedes (287–212 BCE).

In the philosophical works of the Chunqiu and the Zhanguo period, one can find statements containing scientific and technological content as well. This is especially true of the work "Mojing" by Mo Di, which dates from about the same time as "Kao Gong Ji". Nevertheless, "Kao Gong Ji" is the first book containing purely technological content.

Western research on the history of science and technology was for a long time essentially restricted to European civilisation. However, in the last decades a change took place, especially through the pioneering research work on the history of Chinese science and civilisation conducted by English biochemist and science historian Joseph Needham (1900–1995) and his colleagues in China and the West. Interest clearly shifted to civilisations outside of Europe, especially Chinese civilisation. Needham's research work explored the development of the national sciences to construct a world science and trace the connections between the national sciences.

Due to "Kao Gong Ji's" pre-eminent significance for the technologies developed in China, we have decided to present this book to a worldwide audience of readers interested in the history of science and technology, and to offer three translations: modern Chinese and, for the first time, English and German. For the translation, we built on the longstanding, praiseworthy research of Professor Wenren Jun on "Kao Gong Ji", and we feel deeply indebted to him. We also wholeheartedly thank Dr. Matthew Klopfstein and Dr. Marieke Krijnen, who finished the English translation. Further, we thank Mr. Björn Wysfeld who photoshopped the illustrations in this book.

However, the present translations into modern Chinese, English and German do not uncritically follow existing translations. Rather, based on our research, we provided new interpretations of the original text in different places. This is the case for instance in the chapter on the manufacture of carriages, regarding the sizes of the wheels and the hubs; in the chapter on bronze technology, when discussing the reasons for the compliance and non-compliance of the composition of excavated artefacts' bronze with the mixture ratios of the bronzes indicated in "Kao Gong Ji"; and for the design of the ancestor temple in the king's city. We hope that future archaeological findings will clarify remaining questions on the interpretation of "Kao Gong Ji".

May this edition contribute to propagating knowledge about ancient China's scientific achievements all over the world.

Guan Zengjian and Konrad Herrmann
Shanghai 2018

Introduction

1 On Historical Editions and Commenting of "Kao Gong Ji"

The work "Kao Gong Ji" was written in the beginning of the Zhanguo period. After the burning of books on the command of emperor Qin Shihuangdi 秦始皇帝 in the year 213 BCE, which destroyed mostly political, philosophical and literary works, traces of "Kao Gong Ji" disappear. However, the book turned up again during the Western Han 西汉 dynasty. Bibliophile Xian King De of Hejian 河间献王德 obtained it for his collection. At this time, six paragraphs of the book were already missing, namely: the manufacture of metallic agricultural tools, the manufacture of sacrificial garments, the manufacture of leather clothing, the dyeing of clothes, the manufacture of combs and the manufacture of bone products. At this time, another book, "Zhou Li" 周礼 on the rites of the Zhou dynasty, whose chapters were divided according to the designations of six officials, had its sixth chapter "The winter official" (Dong Guan 冬官) missing. The king substituted this missing chapter with the book "Kao Gong Ji", which thus received the head title "The winter official". The title "Kao Gong Ji" (The Artificer's Records) is possibly derived from the Office of Manufactures (*kao gong shi* 考工室), which supervised the manufacture of weapons and armaments during the Han dynasty.

As early as the time of the Han 汉 dynasty, several scholars, among them Ma Rong 马融 (79–166) wrote commentaries on the *Rites of Zhou*, which always contained commentaries on the "Kao Gong Ji" as well. The most influential commentary from this time came from Ma Rong's pupil, the scholar Zheng Xuan 郑玄 (127–200), who endeavoured to commenting all Confucian classics in order to derive a coherent doctrine.

In the year 837, during the Tang 唐 dynasty in Xi'an 西安, twelve Confucian classics were engraved on stone plates, among them the "Kao Gong Ji". This work is named "Kaicheng Shi Jing" 开成石经 (*Stone Classics from the Era Kaicheng*). These stone plates are now kept in the Stele Forest Museum of Xi'an 西安碑林博物馆. They represent the oldest preserved version of "Kao Gong Ji". In 932 during the later Tang dynasty, nine Confucian classics, among which the "Kao Gong Ji", were carved on wooden plates and printed at the initiative of the minister Feng Dao 冯道. However, no copies of this edition survived. The oldest preserved printed version of "Kao Gong Ji" comes from the Southern Song 南宋 dynasty and is kept in the Beijing Library.

 | DOI:10.1163/9789004416949_002

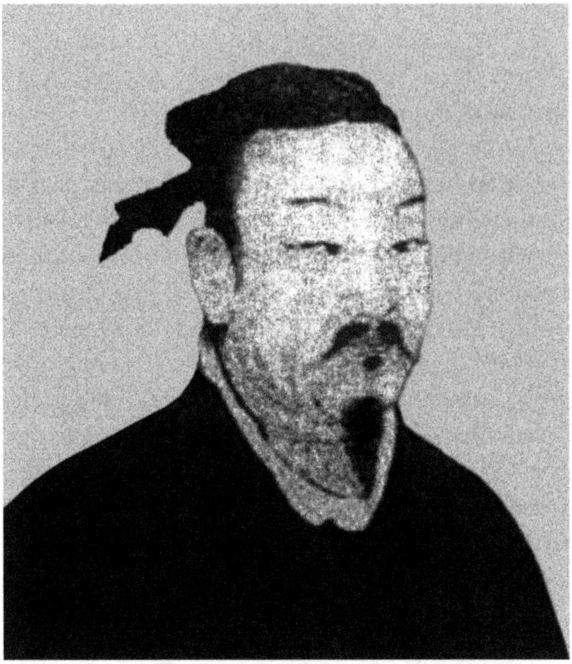

FIGURE 1 Scholar and commentator of "Kao Gong Ji" Zheng Xuan 郑玄 (127–200)
Source: www.Baidu.com, article Zheng Xuan (searched on 8.11.2019)

During the dynasties Song 宋, Ming 明 and Qing 清, many commentaries exclusively dedicated to the work "Kao Gong Ji" appeared. Among the commentaries from the time of the Qing dynasty, the work "Kao Gong Ji Tu" 考工记图 (*The Illustrated Kao Gong Ji*) by Dai Zhen 戴震 (1724–1777), stands out.[1] Dai Zhen was a scholar and collaborator in the compilation of the largest preserved collection of Chinese literature, "Si Ku Quan Shu" 四库全书 (*Complete Book Collection in Four Categories*). His work first appeared in the year 1748. The edition from 1755 is introduced by a praising foreword of Ji Yun 纪昀 (1725–1805), the famous editor of the book collection "Si Ku Quan Shu" 四库全书.

As the title already indicates, Dai Zhen added many knowledgeable illustrations to the "Kao Gong Ji", which significantly facilitate the understanding of the text. Figs. 3 and 4 show a page of this work and an example of an illustration, respectively.

1 Dai Zhen 戴震, *Kao Gong Ji Tu* 考工记图 (Illustrated Kao Gong Ji), Beijing, 1955, Shangwu yinshuguan.

FIGURE 2
Dai Zhen 戴震 (1724–1777), scholar and author of *The Illustrated Kao Gong Ji*

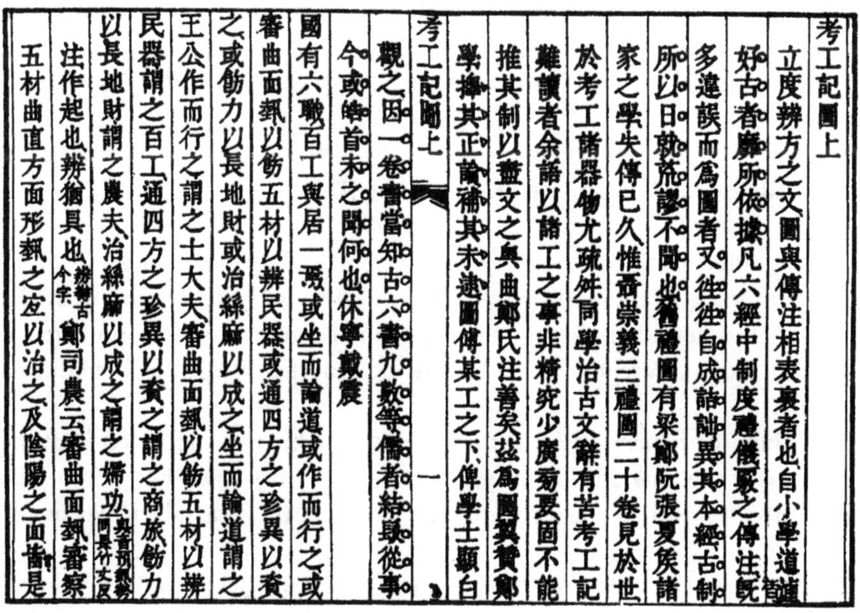

考工記圖上
立度辨方之文圖與傳注相表裏者也自小學道湮
好古者靡所依據凡六經中制度禮儀覈之傳注既
多違誤而為圖者又往往自成詰詘異其本經古制
所以日就荒謬不聞也舊禮圖有梁鄭阮張夏侯諸
家之學失傳已久惟聶崇義三禮圖二十卷見於世
於考工諸器物尤疏舛同學治古文辭有苦考工記
難讀者余語以諸工之事非精究少廣旁要固不能
推其制以盡文之奧曲鄭氏注善矣茲為圖翼贊鄭
學擇其正論補其未逮圖傳某工之下俾學士顯白
考工記圖上 一
觀之因一卷書當知古六書九數等儒者結髮從事
今或皓首未之聞何也休寧戴震
國有六職百工與居一焉或坐而論道或作而行之或
審曲面埶以飭五材以辨民器或通四方之珍異以資
之或飭力以長地財或治絲麻以成之坐而論道謂之
王公作而行之謂之士大夫審曲面埶以飭五材以辨
民器謂之百工通四方之珍異以資之謂之商旅飭力
以長地財謂之農夫治絲麻以成之謂之婦功
注作起也辨猶具也 鄭司農云審曲面埶審察
五材曲直方面形埶之宜以治之及陰陽之面背是

FIGURE 3 A page of Dai Zhen's 戴震 "Kao Gong Ji Tu" 考工记图

FIGURE 4
Illustration of a war chariot in Dai Zhen's "Kao Gong Ji Tu"

Dai Zhen's illustrations have for the largest part been validated by the results of archaeological research since the twentieth century, with some corrections.

Outside of China, the "Kao Gong Ji" was known in Korea since the 10th century, and was printed there in the 15th century with movable types. The book has also been published in Japan since the Ming dynasty.

In Europe, only one translation of the *Rites of Zhou* exists so far: a French translation by French sinologist E. Biot that appeared in the year 1851 under the title *Le Tcheou-Li ou Rites des Tcheou.*[2] In 2013 Wenren Jun published a commented translation into English.[3] Many passages of "Kao Gong Ji" are also quoted in J. Needham's monumental work *Science and Civilisation in China.*[4]

2 Historical Background of "Kao Gong Ji"

As is presented in the following, essential parts of the work "Kao Gong Ji" came into being at the end of the Chunqiu 春秋 and the beginning of the Zhanguo 战国 period, in the principality Qi 齐. Therefore, the comprehension of this book can be greatly assisted if one looks at the history of China during this time.

The Zhou 周 dynasty (11th century–221 BCE), which was founded with the overthrow of the Shang 商 dynasty (ca. 1700–1025 BCE), was a kingdom in the basin of the Yellow River that secured the loyalty of their vassals by bestowing

2 E. Biot, *Le Tcheou-Li ou Rites des Tcheou*, Paris, 1851, reedition: Taipei, Ch'eng Wen Publishing Co., 1975.

3 Wenren Jun, *Ancient Chinese Encyclopaedia of Technology: Translation and Annotation of Kao Gong Ji* (*The Artificer's Record*), London and New York: Routledge 2013

4 J. Needham, *Science and Civilisation in China*, Cambridge University Press. Since 1954, more than twenty volumes have appeared.

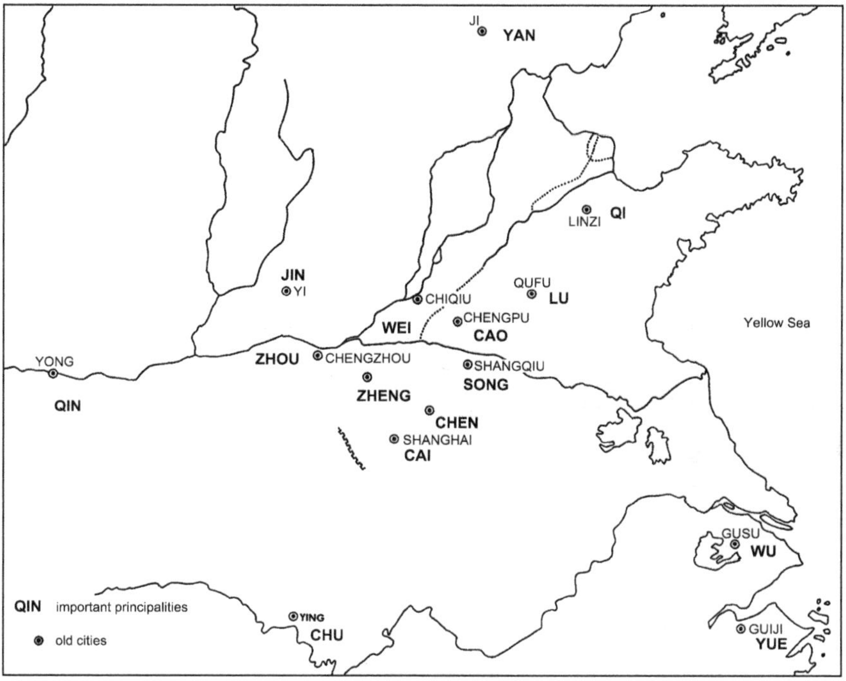

FIGURE 5 China at the time of the Chunqiu 春秋 period
Source: Jacques Gernet, *Die chinesische Welt*, Insel Verlag Frankfurt am Main, 1987, p. 59

fiefs. In the course of time, the central might of the king became ever more weak, and the fiefs of the vassals developed into principalities growing increasingly stronger. In the year 771 BCE, the seat of King You 幽王, situated in the valley of the Wei 渭河 river in the west of the country, was attacked by non-Chinese tribes. Troops from the vassal states Jin 晋 and Qin 秦 could only evacuate the king's clan, and they installed Yi Jiu 宜臼, the successor to the throne, as King Ping 平王 in the new capital nearby Luoyang 洛阳, situated in the east.

The shifting of the capital eastwards marked the beginning of the Eastern Zhou 东周 dynasty (770–256 BCE). At the same time, the period from 770 until 476 BCE is named the Chunqiu 春秋 period, because its history was written down in the annals of the principality Lu 鲁, which bear the name Chunqiu 春秋 (spring and autumn). The following period, which preceded the unification of the empire by the Qin 秦 dynasty in the year 221 BCE, is called the Zhanguo 战国 period, or the Warring States period (475–221 BCE).

Although the royal dynasty exerted only little influence, it acted as symbol of rule and was the binding link between the concurrent principalities (see Fig. 5).

FIGURE 6 The minister Guan Zhong 管仲
Source: www.Baidu.com, article Guan Zhong (searched on 8.11.2019)

During the rule of the duke Huan of Qi 齐桓公 (685–643 BCE), the principality Qi 齐, which was situated in the northern part of the Shandong 山东 peninsula, became the hegemon among the Chinese principalities. As we will see, the work of the duke's minister Guan Zhong 管仲 (ca. 723–645 BCE) formed a crucial contribution to this.

Guan Zhong 管仲 introduced an efficient system of government for the principality, which united civil and military functions. The development of fishing, of the production of salt at the seaside and of textile manufacture and metal processing increased the economic power of the Qi 齐 principality. Duke Huan 桓公 engaged his country in military actions in order to fend off attacks of non-Chinese tribes on neighbouring states. Due to this, Zhou King Hui 周惠王 appointed him as hegemon (*ba* 霸). He had to implement this hegemony not only through military power, but also through restoring the authority of the Zhou 周 king. In several instances, duke Huan 桓公 dispatched troops when neighbouring states in the north were threatened by non-Chinese tribes. In the south, he curbed the threat posed by the state Chu 楚, whose ruler proclaimed himself as king and in this way called into question the reputation of the Zhou 周 king. These successful interventions underpinned the duke Huan of Qi's 齐桓公 prestige as hegemon. On dukes' meetings he convened, he constructed an order of an interstate community of the principalities. However, after the death of duke

Huan 桓公, the hegemony of the principality Qi 齐 was lost. It was seized by duke Wen of Jin 晋文公 (ruled 636–628 BCE) after a struggle for power.

It is highly probable that the work "Kao Gong Ji", which mainly described technologies and crafts oriented at the needs of the ruling class, came into being at this time.

Based on empirical knowledge, "Kao Gong Ji", among other works, laid the basis for the manufacture of war carriages, bronze technology, ceramic products and the manufacture of bows. The book contains the first-known descriptions of the different mixing ratios of copper and tin required for different groups of bronze products according to their functional demands. "Kao Gong Ji" also describes the structure of a combined measurement standard for length, volume and mass, the principles of which were used about 500 years later (during the reign of Wang Mang) to create the combined measurement standard "Jialiang". The original can be admired in the National Museum in Taipei (Taiwan). The measurement standard described in "Kao Gong Ji" is the first in the world to represent a combination of several measurement quantities. "Kao Gong Ji" also specifies a classification of creatures, in which it is remarkable that human beings are considered as part of the fauna. This book also presents for the first time different types of ceramic vessels that were used at the time. The rules for the building of cities laid down in "Kao Gong Ji" have been applied in the construction of the capital Beijing and in other cities. These rules of city building aimed at guaranteeing social order according to the requirements of traditional society, while enabling trade and transport at the same.

For all the products described in "Kao Gong Ji", qualitative data on the kinds of materials needed, and quantitative data on their sizes, volume and mass are attached, and the manufacturing process is detailed. In many places moreover, instructions on how to test the products are provided. Consequently, this technological book laid the foundations for standardization, metrology and quality assurance all at the same time. When the emperor Qin Shihuangdi unified China in the year 221 BCE and then issued a great number of edicts for the unification of the script, weights, measures and the width of wheel tracks, a long tradition in the fields of standardization, metrology and quality assurance already existed in Chinese society.

During the Chunqiu 春秋 period, the economy relied mainly on agriculture. At that time, the relationship between peasants and the ruler was regulated on the basis of the so-called well-field-system, which is also mentioned in the "Kao Gong Ji". This name is derived from the picture of the character for well, "*jing*" 井. According to this system, the fields of each nine peasant families were arranged around one central field. The fruits of the central field had to be delivered to the ruler. This meant a taxation level of 1/9, or 11%.

In the capitals of the principalities, workshops for the manufacture of ceramics and bronze articles existed that had already attained a high level of technology in the Chunqiu 春秋 period. Moreover, at the end of this period, iron metallurgy was developed, probably on the basis of bronze metallurgy. This former technology is not described in the "Kao Gong Ji", most likely because compared with bronze metallurgy, it had not yet ripened and did not serve to satisfy the luxury needs of the ruling class. On the other hand, iron products were possibly described in the lost chapter on agricultural tools.

Officials were mostly recruited from the aristocrats, but there are many examples of officials whose origins remain unclear. Intellectual capabilities and moral integrity were expected from the officials. The education of the Zhou 周 aristocrats comprised six subjects: rites, music, archery, charioteering, writing and calculating.

In the Zhanguo 战国 period, seven powers (*qi xiong* 七雄) emerged: The states Han 韩, Wei 魏 and Zhao 赵, which materialised from the division of the

FIGURE 7 China at the time of the Zhanguo 战国 period
Source: Patricia Buckley Ebrey, *Cambridge Illustrated History China*, Campus Verlag Frankfurt/New York, 1996, p. 40

principality Jin 晋; the state Qi 齐; and furthermore Yan 燕 (in today's Hebei 河北), Qin 秦 (in today's Shaanxi 陕西) and Chu 楚 (in the middle reaches of the Yangtse 长江 river) (see Fig. 7). In contrast to the Chunqiu 春秋 period, when the principalities were more or less city states, these states comprised a territory that had to be defended.

Whereas wars in the Chunqiu 春秋 period were still a matter for the aristocrats, in the Zhanguo 战国 period that followed, ever more resources were used for warfare. Now, using a strong infantry recruited from peasants, siege wars were waged. Mobile cavalry gradually replaced the war chariots. The conduct of war increasingly depended on economic resources. This resulted in rulers' strong interest in economic questions, and the Zhanguo 战国 period is characterised by a flourishing economy. Large irrigation facilities were built for agriculture, and the introduction of crop rotation led to more harvests. The rulers promoted the opening of virgin soil in order to attract settlers.

In the fourth and third century BCE, iron tools were introduced in agriculture. Contrary to those used in the Roman Empire, these tools were not forged but cast which enabled a very efficient manufacturing process. Another important innovation was related to horse carriages. The introduction of shafts with breast-strap harnessing enabled a more effective use of the horses' power, since the throat-and-girth yoke used previously hindered the breathing of the animals. In this way, a horse could pull a load that previously required two horses.

While trade in the Chunqiu 春秋 period was restricted to luxury goods, in the Zhanguo 战国 period, an ever broader trade in diverse consumer goods can be observed. Merchants began to form a class of their own, a development also mentioned in "Kao Gong Ji". The cities, which were formerly merely the seat of the ruler, developed into centres of craft and trade. The introduction of money in the late Chunqiu 春秋 period, in the form of cast bronze coins, is connected to this newly emerging trade as well.

The incessant wars between the principalities also led to the decline of old aristocrat families. Meanwhile, until the year 323 BCE, the rulers of the remaining states assumed the title "king" (*wang* 王), indicating the total decline of Zhou 周 rule. Parallel to the decline of old aristocrat families, who ruled with the help of unwritten rites, bodies of law were gradually created on the basis of which civil servants governed the population. These officials received wages that were measured according to prescribed quantities of grain. Moreover, prominent dignitaries received (non-inheritable) fiefs with a certain number of households on which they could impose taxes.

Beginning in 350 BCE, the state Qin 秦, situated in the West of China, first defeated the principalities Shu 蜀 and Ba 巴 located in today's Sichuan 四川,

FIGURE 8
Emperor Qin Shihuangdi 秦始皇帝
Source: W. Krüger, *Shihuang Di und der Alltag im alten China*, ArteMedia Köln 1981, p. 6

and then defeated the principalities along the Yellow River in succession. The strength of the state Qin 秦 can be explained by an improved food supply due to extensive irrigation systems, the large-scale manufacture of iron weapons and the effective administration of the state by permanent civil servants. In an unbroken triumphal procession between 230 and 221 BCE, Qin 秦 conquered the principalities Han 韩, Wei 魏, Chu 楚, Zhao 赵, Yan 燕 and Qi 齐, one after the other. In the year 221 BCE, King Zheng of Qin 秦王政 proclaimed himself as "First Emperor of Qin" (Qin Shihuangdi 秦始皇帝) and governed a unified China.

The cultural achievements of the Zhanguo 战国 period include the various philosophy schools. The most important of these, due to their lasting impact, are the Confucianists and the Daoists. Kongzi 孔子 (551–479 BCE) who the Jesuits called Confucius, a latinised form of Kong Fu Zi 孔夫子, founded a school whose pupils became successful officials, political advisers in the courts of the principalities, rites masters and teachers themselves.

Confucius himself, who spent a long time searching for a ruler who could realise his ethical teachings, only held rather low, temporary posts. Politically, he strived for the restauration of the Zhou 周 dynasty's power as it had existed in its early years. He combined this with strict requirements for rulers' morals. They should provide the people with an example of decency. Order in society

FIGURE 9
Philosopher Kong Zi 孔子

should rely on relationships of subordination: the servant must be subordinate to the ruler, the son to the father, the younger to the elder brother and the woman to the man. A famous representative of the Confucianists is Meng Zi 孟子 (ca. 382–300 BCE).

The school of the Daoists was founded by Laozi 老子, about whom almost nothing is known for certain. It is said that he was a contemporary of Confucius. The philosopher Zhuang Zi 庄子 (ca. 365–280 BCE), who continued the teachings of the Dao 道, polemicised against the Confucianists in a polished, literary manner. Dao literally means "way" and further "world principle". The Daoists criticised what they viewed as undesirable trends in society caused by the greed and warmongering of the rulers, ever more laws whose consequences could not be foreseen and complicated rites. The ideal of the Daoists was a wise ruler who would govern the people discreetly and mercifully.

The Yin-Yang 阴阳 school and the followers of the theory of the five phases of transformation (*wu xing* 五行) clearly trace back to fortune-telling traditions. These teachings laid the foundation for a scientific view of the world, as is visible in the "Yijing" 易经, the *Book of Changes*. Echoes of these teachings can also be found in the work "Kao Gong Ji".

Another school represented the philosophy of mohism, and was founded by Mo Di 墨翟 (ca. 480–390 BCE). Mo Di 墨翟 attacked the Confucianists as

FIGURE 10
Philosopher Meng Zi 孟子
Source: *Mong Dsi*, Eugen Diederichs Verlag Jena 1916, frontispiece

FIGURE 11
Philosopher Zhuang Zi 庄子
Source: *Dschuang Dsi, Das wahre Buch vom südlichen Blütenland*, Eugen Diederichs Verlag München, 1988, frontispiece

well, and these were henceforth called "*ru* 儒", among other things, because of their cultivation of expensive funeral ceremonies (which was one of their sources of income). Mo Di 墨翟 demanded universal human love; he strictly rejected wars, although he did think of how one can defend oneself effectively. Therefore, the Mohists offered the dukes their service as advisers in technical questions of siege war.

Successful reformers of the administration and economy, like Shang Yang 商鞅 (390?–338 BCE), pursued a strict application of laws, laying down precise punishments and rewards. These bodies of laws gradually replaced the unwritten rites of the aristocrats. Han Fei Zi 韩非子 (ca. 280–233 BCE) in particular delivered the philosophical foundation of this school of thought. Since the Han 汉 dynasty, the followers of Han Fei Zi were called Legists (*fa jia* 法家). Impoverished aristocrats offered their services to the dukes, acting as advisers, and were also called "guests" (*binke* 宾客) because they lived in the court permanently.

Minister Lü Buwei 吕布韦, who served the later Qin emperor Qin Shihuangdi 秦始皇帝 and ruled from 250 until 238 BCE, attracted many such guests and scholars. He wrote an eclectical philosophical work titled *Spring and Autumn of Lü Buwei* ("Lü Shi Chun Qiu" 吕氏春秋).

FIGURE 12
Minister Lü Buwei 吕布韦
Source: www.Baidu.com, article Lü Buwei (searched on 8.11.2019)

3 Location and Period of Origin of "Kao Gong Ji"

The location and period of origin of "Kao Gong Ji" can be determined based on certain features of the text and on the similarities between excavated artefacts and their descriptions in this work.

Scholar and poet Guo Moruo 郭沫若 (1892–1978), who later acted as president of the Academia Sinica, analysed the text of "Kao Gong Ji" and concluded that it represents a book on the officialdom of the principality Qi 齐.[5] As proof he puts forward that the principality Qi is not mentioned in the text, contrary to the other principalities. The text contains six idiomatic expressions from Qi, and the book uses the system of measurements of the principality Qi.

Xuan Zhaoqi 宣兆琦[6] further mentions that the detailed descriptions of bronze technology correspond with archaeological discoveries in the area of the former principality Qi. The mentioning of workmen in the third place of rank, after the princes and officials, reflects the economic development of this region, which is also presented in the work "Guan Zi" 管子. Finally, excavations have confirmed the description of carriage wheels as having thirty spokes.

As for its time period of origin, "Kao Gong Ji" is generally assumed to have originated in the late years of the Chunqiu 春秋 and the early years of the Zhanguo 战国 periods. A good reason for this is that the bronze alloys mentioned in the work can only be confirmed since the Chunqiu period. Moreover, the various principalities mentioned in the book existed only since this period as well. The classes listed in "Kao Gong Ji" and the well-field-system correspond to the reforms of Guan Zhong 管仲 (723 or 716–645 BCE), so that large parts of the book could have come into being during the rule of prince Huan of Qi 齐桓公. During his rule, Guan Zhong carried out economic reforms that turned the principality Qi into the strongest power in the region. Because in the history chronicle "Shi Ji" 史记 (*Historical Records*) of Sima Qian 司马迁 (145 or 135–87? BCE) a certain Chen Wan 陈完 is mentioned, who was the chief superintendent of the workmen, Xuan Zhaoqi 宣兆琦 assumes that the "Kao Gong Ji" was written by Chen Wan.

On the other hand, the "Kao Gong Ji" also shows linguistic dissimilarities that make it likely that some sections of the book came into being at a later

5 Guo Moruo 郭沫若, *Shi pipan shu* 十批判书 (Book of ten criticisms), Xin wenyi chubanshe 1951, p. 30.

6 Xuan Zhaoqi 宣兆琦, *"Kao Gong Ji" de guobie he chengshu niandai* 考工记的国别和成书年代 (Location and time of the origin of "Kao Gong Ji"), Ziran kexueshi yanjiu, 12 (1993) 4, pp. 297–303.

time. For instance, the descriptions about the planning of cities have not been confirmed by excavations of cities from the Chunqiu and Zhanguo periods. Therefore, this section was probably written only during the Qin 秦 dynasty (221–206 BCE) or the Western Han 西汉 dynasty (206 BCE–24 CE).[7]

Critics of Guo Moruo's above-mentioned thesis (that "Kao Gong Ji" is a book of the officialdom of the principality Qi) argue that features of certain excavated bronze artefacts do not correspond with the data in "Kao Gong Ji". This applies for instance to the sizes of some bronze clocks and weapons found in the area of the former principality Qi.[8] However, it must be remarked that from seven investigated sizes of bronze clocks, three do not correspond to the data provided in "Kao Gong Ji", but four correspond well. Discrepancies between the sizes of spears found and sizes of those discussed in "Kao Gong Ji" can be attributed to the fact that the found spears did not concern ceremonial weapons, but were intended for battle. This also applied to the six mixing ratios for bronze alloys.

During the Han dynasty, the book "Kao Gong Ji" was incorporated by King De of Hejian 河间德王 (died 130 BCE), a great collector of old books, into the *Rites of Zhou* ("Zhou Li") 周礼, which describes the rites of the Zhou dynasty. Even though in terms of content "Kao Gong Ji" has nothing to do with the rites of the Zhou dynasty, it seems to be that we have King De of Hejian to thank for making sure that the "Kao Gong Ji" came down to us.

On the whole, one can establish that the thesis situating the origins of "Kao Gong Ji" in the principality Qi and its largest parts in the Chunqiu and the Zhanguo periods is very probable. Perhaps the results of further excavations will underpin the knowledge about this issue further.

---

7 Song Huan 宋烜, *"Kao Gong Ji. Jiangren" chengshu niandai xi* 考工记匠人成书年代析 (Analysis of the formation time of the section Building Workers in "Kao Gong Ji"), Nanfang wenwu, (1998), 2, pp. 98–100.

8 Liu Guangding 刘广定, *"Kao Gong Ji" fei Qi guo guan shu zhi zheng* 考工记非齐国官书之证 (Disapproval of Kao Gong Ji as an official document from Qi State), Di ba jie kexueshi yantaohui huikan (Proceedings of 8th seminar on history of science), Zhongyang kejiuyuan kexueshi weiyuanhui, Taibei, 2008, pp. 1–7.

PART 1

The Text

卷上 (Part One)

AC 总叙

国有六职，百工与居一焉。或坐而论道；或作而行之；或审曲面势，以饬五材，以辨民器；或通四方之珍异以资之；或饬力以长地财；或治丝麻以成之。坐而论道，谓之王公；作而行之，谓之士大夫；审曲面势，以饬五材，以辨民器，谓之百工；通四方之珍异以资之，谓之商旅；饬力以长地财，谓之农夫；治丝麻以成之，谓之妇功。

粤无鎛，燕无函，秦无庐，胡无弓车。粤之无鎛也，非无鎛也，夫人而能为鎛也；燕之无函也，非无函也，夫人而能为函也；秦之无庐也，非无庐也，夫人而能为庐也；胡之无弓车也，非无弓车也，夫人而能为弓车也。知者创物，巧者述之守之，世谓之工。百工之事，皆圣人之作也。烁金以为刃，凝土以为器，作车以行陆，作舟行水，此皆圣人之所作也。

天有时，地有气，材有美，工有巧，合此四者，然后可以为良。材美工巧，然而不良，则不时，不得地气也。橘踰淮而北为枳，鸜鵒不踰济，貉逾汶则死，此地气然也；郑之刀，宋之斤，鲁之削，吴粤之剑，迁乎其地而弗能为良，地气然也。燕之角，荆之干，妢胡之笴，吴粤之金锡，此材之美者也。天有时以生，有时以杀；草木有时以生，有时以死；石有时以泐；水有时以凝，有时以泽:此天时也。

凡攻木之工七，攻金之工六，攻皮之工五，设色之工五，刮摩之工五，抟埴之工二。攻木之工:轮、舆、弓、庐、匠、车、梓；攻金之工:筑、冶、凫、栗、段、桃；攻皮之工:函、鲍、韗、韦、裘；设色之工:画、缋、锺、筐、㡛；刮摩之工:玉、楖、雕、矢、磬；抟埴之工:陶、旊。有虞氏上陶，夏后氏上匠，殷人上梓，周人上舆。故一器而工聚焉者，车为多。

MC 总叙

一个国家的内部分工，有六种职事，百工是其中之一。在国家的人群中，有坐而论道的；有制订政令努力推行的；有审视材料外形考究其内在质地，以整饬五材[1]为民众制作各种器物的；[2]有奔走各地买卖珍宝异物以互通有无的；有辛苦劳作从事农耕以开发土地财富的；有整治丝麻制作衣被的。坐而论道的，称为王公；制订政令努力推行的，称为士大夫；审视材料制作器物的，称作百工；奔走各地买卖珍宝异物以互通有无的，称为商

1 “五材”，根据汉代郑玄的理解，指金（铜）、木、皮、玉、土。

2 “辨”，古语通“办”，此处为治理、整备之意。

 | DOI:10.1163/9789004416949_003

旅；辛苦劳作从事农耕以开发土地财富的，称为农夫；整治丝麻制作衣被的，称为妇功。

粤[3]地没有专门制作鎛[4]的工匠，燕地[5]没有专门制作函[6]的工匠，秦地[7]没有专门制作庐器[8]的工匠，胡地[9]没有专门制作弓车的工匠。粤地没有专门制作鎛的工匠，并非那里的人们不懂如何制鎛，而是人人都会的缘故。燕地没有专门制作函的工匠，并非那里的人们不懂如何制函，而是人人都会的缘故。秦地没有专门制作庐器的工匠，并非那里的人们不懂如何制庐器，而是人人都会的缘故。胡地没有专门制作弓车的工匠，并非并非那里的人们不懂如何制弓车，也是因为人人都会制作的缘故。有大智慧的人发明器物，心灵手巧者将其记录下来，世代传承，人们把这样的人叫做工匠。百工之事，都是圣人的创作。熔炼金属制作兵刃工具，陶制泥土以成器物，制作车辆陆地行走，造就舟船水域通行，这些，都是圣人的发明。

Part 1: Introduction*

The country has six classes, one of which is called the hundred artificers. While some discuss politics sitting quietly, others try hard to enforce it. Again others examine round and even surfaces and other shapes in order to process the five materials[10] from which goods for the people are manufactured. Some trade with precious things from the four corners of the world and amass them. Others till the land with the force of their body to grow grain, or they get silk and hemp to produce cloth from. Those who sit quietly discussing politics are the kings and princes. Those who try hard to enforce the politics are the officials. Those who examine round and even surfaces and other shapes in order to process the five materials and to manufacture goods for the people are the hundred artificers. Those who trade with precious things from the four corners of the world and amass them are the merchants. Those who till the land with

3 这里的“粤”通“越”，不是指现在的广东，而是指先秦时的越国，在今浙江一带。

4 “鎛”，一种青铜工具，锄铲之类。

5 “燕”，指当时的燕国，在今河北北部和辽宁一带。

6 “函”，这里指皮甲或铠甲之类。

7 “秦”，指秦国，当时秦国位于现在陕西甘肃一带。

8 “庐”，指当时通行的长兵器如戈、矛、戟等的木、竹柄。

9 “胡”，这里指当时我国北方和西北方的少数民族。

* See the commentaries: On the History of Edition and Commenting of *Kao Gong Ji*, Historical Background of *Kao Gong Ji* and Location and Time of Origin of *Kao Gong Ji*.

10 *Five materials* – the five materials metal, wood, leather, jade and earth are meant.

the force of their body are the peasants and those who get silk and hemp to produce cloth from are the weaving women.

The principality Yue 粤[11] has no artificers who are specialised in manufacturing cast hoes. The principality Yan 燕[12] has no artificers who are specialised in manufacturing harnesses. The principality Qin 秦[13] has no artificers who are specialised in manufacturing lances, and the neighbouring people have no artificers who are specialised in manufacturing bows and carriages. If one says the principality Yue 粤 has no artificers who are specialised in manufacturing cast hoes, it does not mean that there are no hoes; on the contrary, every workman is able to manufacture hoes. If one says the principality Yan 燕 has no artificers who are specialised in manufacturing harnesses, it does not mean that there are no harnesses; on the contrary, every workman is able to manufacture harnesses. If one says the principality Qin 秦 has no artificers who are specialised in manufacturing lances, it does not mean that there are no lances; on the contrary, every workman is able to manufacture lances. If one says the neighbouring people have no artificers who are specialised in manufacturing bows and carriages, it does not mean that there are no bows and carriages; on the contrary, every workman is able to manufacture bows and carriages. Talented men produce things; dextrous men pass on their skills and keep them for the coming generations. These are called artificers. The things manufactured by the hundred artificers are all inventions by the sages. They melted metals to make sharp blades from. They burnt earth to make vessels from. They built carriages to go over the land. They built ships to sail on the sea. All this was created by the sages.

Heaven has its seasons, earth has its Qi 气,[14] materials have their beauty, artificers have their skills. If these four conditions come together, then excellent things can be formed. But if the material was good and the artificer was skilled, and nonetheless no good product was created, then the season was not right and one did not get the Qi 气 of the earth. If one plants mandarin trees north

11 *Principality Yue* 粤 – principality in the territory of today's province Zhejiang 浙江. The capital Ji 稽 lies near today's Shaoxing 绍兴.

12 *Principality Yan* 燕 – principality in the territory of today's provinces Hebei 河北 and Liaoning 辽宁. The capital Ji 蓟 was situated near today's capital Beijing 北京.

13 *Principality Qin* 秦 – principality in western China in the western Shaanxi 陕西 and south-eastern Gansu 甘肃 provinces. The capital Yong 雍 lies near the city Fengxiang 凤翔 in Shaanxi 陕西. Archaeological findings show that the manufacture of tools and weapons from bronze and iron developed early here.

14 *Qi* 气 – Qi 气 is a term of the original natural sciences that can be interpreted as the flow of energy in the earth and in the air, or as a vapour. Often, the effect of Qi 气 is connected with magical ideas.

of the Huai 淮 river,[15] they change into the three-leaved tangerine. The crested myna never flies north beyond the Ji 济 river,[16] the marten dies if he crosses the Wen 汶 river[17] to the south. All this is caused by the Qi 气 of the earth!

The knives from the principality Zheng 郑,[18] the axes from the principality Song 宋,[19] the scraping knives from the principality Lu 鲁[20] with which the script on bamboo strips is corrected, the swords from the principalities Wu 吴 and Yue 粤 – all these are excellent products. They would not be so excellent if they are not made in these regions. This is also caused by the Qi 气 of the earth! The ox horns from the principality Yan 燕, the bows from Jing 荆,[21] the arrow shafts from Fenhu 妢胡[22] and copper and tin from the principalities Wu 吴[23] and Yue 粤 count as top-quality materials. Nature has its time to grow and its time to die. Grasses and trees have their time to grow and their time to die. Rocks have their time when they develop crevices. The water has its time to freeze and it has its time to melt. All this is caused by the seasons of heaven!

There are seven different crafts that process wood, six different crafts that process metals, five different crafts that process leather, five different crafts that process dyes, five different crafts that grind and polish and two different crafts that produce ceramics. The wood processing crafts comprise the manufacture of wheels, carriage bodies, bows, lances, carpenter goods, chariots and stands for musical instruments. The metal processing crafts comprise the manufacture of script knives, arrowheads, bells, weights, farm tools and forged swords. The leather processing crafts comprise leather tanning, drums, knee pads for mourning robes and furs. The dye processing crafts comprise the painting of coloured patterns, the embroidering of coloured patterns, the dyeing of

---

15 *Huai 淮 river* – The Huai 淮 river originates in Henan 河南 and flows eastwards through the provinces Anhui 安徽 and Jiangsu 江苏.

16 *Ji 济 river* – describes a side arm of the Huanghe 黄河 (Yellow River) in the provinces Henan 河南 and Shandong 山东.

17 *Wen 汶 river* – another name for the Changjiang 长江 or Yangtse River.

18 *Principality Zheng 郑* – principality in the territory of Eastern Shaanxi 陕西 province.

19 *Principality Song 宋* – principality in the border region of the provinces Henan 河南, Shandong 山东, Jiangsu 江苏 and Anhui 安徽.

20 *Principality Lu 鲁* – principality in the south-western part of today's Shandong 山东 province. The capital was in Qufu 曲阜, the birthplace of Confucius.

21 *Jing 荆* – a city belonging to the principality Chu 楚 on the territory of today's city Hengyang 衡阳.

22 *Fenhu 妢胡* – an old geographic name that could not be identified reliably. It was known as a region where plenty of bamboo grew.

23 *Principality Wu 吴* – principality in the border region of the provinces Jiangsu 江苏, Anhui 安徽 and Zhejiang 浙江. The principalities Wu 吴 and Yue 越 were famous for their swords.

feathers, the dyeing of cloth and the dyeing of silk. Crafts for grinding and polishing comprise the manufacture of jade objects, combs, bone carvings, arrows and musical instruments made of stone. Crafts that process ceramics comprise the making of ceramic dishes and ceramic containers.

The official who supervised the manufacture of ceramics was highly regarded by the ruler You Yu Shi 有虞氏.[24] The rulers of the Xia 夏 dynasty[25] held the official who was responsible for hydraulic and civil engineering in high regard. The people of the Yin 殷 dynasty[26] held the official who was responsible for the manufacture of ritual musical instruments in high regard. The people of the Zhou 周 dynasty[27] held the official who was responsible for the manufacture of chariots in high regard.

Teil 1: Einführung*

Im Land gibt es sechs Stände, unter denen die hundert Handwerker einen bilden. Die einen sitzen still und erörtern die Politik, während andere sich mühen, um sie durchzusetzen. Wieder andere prüfen runde und ebene Flächen und andere Formen, um die fünf Materialien[28] zu bearbeiten, aus denen Produkte für das Volk hergestellt werden. Manche handeln mit kostbaren Dingen aus den vier Weltgegenden und häufen sie auf. Andere bearbeiten mit der Kraft ihres Körpers das Land, um Getreide anzubauen, oder sie gewinnen Seide und Hanf, um daraus Stoffe zu weben. Die still sitzen und die Politik erörtern, sind die Könige und Fürsten. Die sich mühen, um die Politik durchzusetzen, sind die Beamten. Diejenigen, die runde und ebene Flächen und andere Formen prüfen, um die fünf Materialien zu bearbeiten und daraus

24 *You Yu Shi* 有虞氏 – You Yu Shi 有虞氏 is the name of a tribe. Its tribal chief Yu Shun 虞舜 belonged to the legendary rulers. It is said that he invented the calendar and weights and measures and introduced more lenient punishments.

25 *Xia* 夏 *dynasty* – The Xia 夏 dynasty is the oldest known dynasty. It lasted approximately from the 21st until the 16th century BCE.

26 *Yin* 殷 *dynasty* – The Yin 殷 dynasty, which is also called Shang 商 dynasty, lasted approximately from the 16th until the 11th century BCE. Yin was the name of its capital.

27 *Zhou* 周 *dynasty* – The Zhou 周 dynasty lasted from approximately the 11th century until 221 BCE. The Zhou 周 dynasty is divided into the Western Zhou 西周 dynasty (ca. 11th century–770 BCE), the Chunqiu 春秋 period (770–476 BCE) and the Zhanguo 战国 period (475–221 BCE).

* Siehe Kommentare: On the history of editing and commenting the "Kao Gong Ji", Historical background of "Kao Gong Ji" und Location and period of origin of "Kao Gong Ji".

28 *fünf Materialien* – Unter den fünf Materialien versteht man Metall, Holz, Leder, Jade und Erde.

Produkte für das Volk herstellen, sind die hundert Handwerker. Diejenigen, die mit kostbaren Dingen aus den vier Weltgegenden handeln und sie aufhäufen, sind die Kaufleute. Die mit der Kraft ihres Körpers das Land bearbeiten, sind die Bauern, und die Seide und Hanf gewinnen, um daraus Stoffe herzustellen, sind die Weberinnen.

Im Fürstentum Yue 粤[29] gibt es keine Handwerker, die darauf spezialisiert sind, gegossene Hacken herzustellen. Im Fürstentum Yan 燕[30] gibt es keine Handwerker, die darauf spezialisiert sind, Harnische herzustellen. Im Fürstentum Qin 秦[31] gibt es keine Handwerker, die darauf spezialisiert sind, Lanzen herzustellen, und die Nachbarvölker haben keine Handwerker, die darauf spezialisiert sind, Bögen und Wagen herzustellen. Wenn es heißt, im Fürstentum Yue 粤 gäbe es keine Handwerker, die darauf spezialisiert sind, gegossene Hacken herzustellen, bedeutet das nicht, man hätte dort keine Hacken, vielmehr vermag jeder Handwerker Hacken herzustellen. Wenn es heißt, im Fürstentum Yan 燕 gäbe es keine Handwerker, die darauf spezialisiert sind, Harnische herzustellen, bedeutet das nicht, sie hätten keine Harnische, vielmehr vermag ein jeder Handwerker Harnische herzustellen. Wenn es heißt, im Fürstentum Qin 秦 gäbe es keine Handwerker, die darauf spezialisiert sind, Lanzen herzustellen, bedeutet das nicht, man hätte dort keine Lanzen, vielmehr vermag ein jeder Handwerker Lanzen herzustellen. Wenn es heißt, die Nachbarvölker hätten keine Handwerker, die darauf spezialisiert sind, Bögen und Wagen herzustellen, bedeutet das nicht, sie hätten keine Bögen und Wagen, vielmehr vermag ein jeder Handwerker Bögen und Wagen herzustellen. Talentierte Menschen produzieren Geräte, geschickte Menschen geben ihre Fertigkeiten weiter und bewahren sie. Danach richten sich die Generationen der Handwerker. Die von den hundert Handwerkern hergestellten Dinge sind alle von Heiligen geschaffene Erfindungen. Sie schmolzen Metalle, um daraus scharfe Klingen zu machen. Sie brannten Erde zu Keramikgefäßen. Sie bauten Wagen, um auf dem Land zu fahren, sie bauten Schiffe, um auf dem Wasser zu fahren. Das alles wurde von Heiligen geschaffen.

---

29 *Fürstentum Yue* 粤 – auf dem Territorium der heutigen Provinz Zhejiang 浙江 gelegen. Die Hauptstadt Ji 稽 lag nahe dem heutigen Shaoxing 绍兴.

30 *Fürstentum Yan* 燕 – auf dem Territorium der heutigen Provinzen Hebei 河北 und Liaoning 辽宁 gelegen. Die Hauptstadt Ji 蓟 lag in der Nähe der heutigen Hauptstadt Beijing 北京.

31 *Fürstentum Qin* 秦 – in Westchina im westlichen Teil der Provinz Shaanxi 陕西 und im südöstlichen Teil der Provinz Gansu 甘肃 gelegen. Die Hauptstadt Yong 雍 lag nahe der Stadt Fengxiang 凤翔 in Shaanxi 陕西. Archäologische Funde bestätigten, dass die Herstellung von Werkzeugen und Waffen aus Bronze und Eisen hier früh entwickelt war.

Die Natur hat ihre Jahreszeiten, die Erde hat ihr Qi 气,[32] die Materialien haben ihre Schönheit, die Handwerker haben ihre Geschicklichkeit. Wenn diese vier Bedingungen zusammenkommen, dann können vorzügliche Dinge entstehen. Aber wenn das Material gut und der Handwerker geschickt ist und dennoch kein gutes Produkt entsteht, dann stimmte die Jahreszeit nicht und man erlangte nicht das Qi 气 der Erde. Wenn man Mandarinenbäume nördlich des Huai-Flusses 淮[33] pflanzt, verwandeln sie sich in die dreiblättrige Orange. Der Schopfmaina fliegt niemals nördlich über den Ji-Fluss 済[34] hinaus, der Marderhund überlebt nicht, wenn er nach Süden über den Wen-Fluss 汶[35] setzt. Das bewirkt das Qi 气 der Erde!

Die Messer des Fürstentums Zheng 郑,[36] die Beile des Fürstentums Song 宋,[37] die Schabemesser aus dem Fürstentum Lu 鲁,[38] mit denen man die Schrift auf den Bambustäfelchen korrigiert, die Schwerter aus den Fürstentümern Wu 吴 und Yue 粤 – das alles sind vorzügliche Produkte. Sie können nicht so exquisit sein, wenn sie nicht in diesen Gegenden hergestellt werden. Das bewirkt ebenfalls das Qi 气 der Erde! Die Rinderhörner aus dem Fürstentum Yan 燕, die Bögen aus Jing 荊,[39] die Pfeilschäfte aus Fenhu 妢胡,[40] Kupfer und Zinn aus den Fürstentümern Wu 吴[41] und Yue 粤 zählen zu den erstklassigen Materialien: Die Natur hat ihre Zeit zum Wachsen und ihre Zeit zum Sterben. Die Gräser und Bäume haben ihre Zeit zum Wachsen und ihre Zeit zum Sterben. Die Felsen haben ihre Zeit, in der sie Risse bilden. Das

---

32 *Qi* 气 – Qi 气 ist ein Begriff der ursprünglichen Naturwissenschaften, der als Energiefluss in der Erde und in der Luft oder als Dampf interpretiert werden kann. Oft ist die Wirkung des Qi mit magischen Vorstellungen verknüpft.

33 *Huai-Fluss* 淮 – der Huai 淮-Fluss entspringt in Henan 河南 und fließt nach Osten durch die Provinzen Anhui 安徽 und Jiangsu 江苏, wo er heute in den Yangtse mündet. Der Huai-Fluss hatte im Laufe der Geschichte mehrmals seinen Lauf geändert.

34 *Ji-Fluss* 済 – beschreibt einen Seitenarm des Huanghe 黄河 (Gelben Flusses) in den Provinzen Henan 河南 und Shandong 山东.

35 *Wen-Fluss* 汶 – ein anderer Name für den Yangtse-Fluss, auch Changjiang 长江 genannt.

36 *Fürstentum Zheng* 郑 – auf dem Territorium im Osten der Provinz Shaanxi 陕西 gelegen.

37 *Fürstentum Song* 宋 – Fürstentum auf der Grenze zwischen den Provinzen Henan 河南, Shandong 山东, Jiangsu 江苏 und Anhui 安徽.

38 *Fürstentum Lu* 鲁 – im südwestlichen Teil der heutigen Provinz Shandong 山东 gelegen. Die Hauptstadt war Qufu 曲阜, der Geburtsort von Konfuzius.

39 *Jing* 荆 – eine Stadt, die zum Fürstentum Chu 楚 gehörte, auf dem Territorium der heutigen Stadt Hengyang 衡阳 gelegen.

40 *Fenhu* 妢胡 – ein alter Ortsname, der nicht zuverlässig identifiziert werden konnte. Jedenfalls handelt es sich um eine Region, in der reichlich Bambus wächst.

41 *Fürstentum Wu* 吴 – Fürstentum im Grenzbereich der Provinzen Jiangsu 江苏, Anhui 安徽 und Zhejiang 浙江. Die Fürstentümer Wu 吴 und Yue 越 waren für ihre Schwerter berühmt.

Wasser hat seine Zeit, um zu erstarren, und es hat seine Zeit, um zu schmelzen. Das alles bewirken die Jahreszeiten der Natur!

Es gibt sieben verschiedene Handwerke, die Holz verarbeiten, sechs verschiedene Handwerke, die Metalle verarbeiten, fünf verschiedene Handwerke, die Farben verarbeiten, fünf verschiedene Handwerke, die schleifen und polieren, und zwei verschiedene Handwerke, die Keramik herstellen. Die Holz verarbeitenden Handwerke sind: Rädermacher, Wagenkastenbauer, Bogenmacher, Lanzenmacher, Tischler, Wagenbauer und Tischler für Gestelle zum Aufhängen von Musikinstrumenten. Die Metall verarbeitenden Berufe sind: Hersteller von Schreibmessern, Hersteller von Pfeilspitzen, Glockengießer, Gießer von Gewichtsstücken, Hersteller von landwirtschaftlichen Geräten und Schwertschmiede. Die Leder verarbeitenden Berufe sind: Gerber, Trommelmacher, Schneider von Knieschützern für Trauergewänder und Kürschner. Die Farbe verarbeitenden Berufe sind: Maler farbiger Muster, Stickerinnen farbiger Muster, Färber für Federn, Färber für Stoffe und Seidenfärber. Berufe, bei denen geschliffen und poliert wird, sind: Schleifer von Jadegerätschaften, Kammmacher, Knochenschnitzer, Pfeilmacher, Hersteller von Musikinstrumenten aus Stein. Berufe, die Keramik herstellen, sind: Töpfer von keramischen Essgefäßen und Töpfer für Gefäße zum Aufbewahren von Lebensmitteln. Der Herrscher You Yu Shi 有虞氏[42] schätzte den Beamten hoch, der die Herstellung von Keramik beaufsichtigte. Die Herrscher der Xia-Dynastie 夏[43] schätzten den Beamten hoch, der für den Wasserbau und das Bauwesen verantwortlich war. Die Leute der Yin-Dynastie 殷[44] schätzten den Beamten hoch, der für die Herstellung ritueller Musikinstrumente verantwortlich war. Die Leute der Zhou-Dynastie 周[45] schätzten den Beamten hoch, der für die Herstellung von Streitwagen verantwortlich war.

42 *Herrscher You Yu Shi 有虞氏* – You Yu Shi 有虞氏 ist der Name eines Stammes. Sein Stammeshäuptling Yu Shun 虞舜 zählt zu den legendären Herrschern. Es wird überliefert, dass er den Kalender und die Maße und Gewichte erfunden und mildere Strafen eingeführt hatte.

43 *Xia 夏-Dynastie* – Die Xia 夏-Dynastie ist die älteste bekannte Dynastie. Sie währte etwa vom 21. bis zum 16. Jahrhundert v. Chr.

44 *Yin 殷-Dynastie* – Die Yin 殷-Dynastie, auch Shang 商-Dynastie genannt, währte etwa vom 16. bis zum 11. Jahrhundert v. Chr. Yin 殷 war der Name ihrer Hauptstadt.

45 *Zhou 周-Dynastie* – Die Zhou 周-Dynastie währte etwa vom 11. Jahrhundert bis 221 v. Chr. Die Zhou-Dynastie wird in die Westliche Zhou 西周-Dynastie (etwa 11. Jahrhundert bis 770 v. Chr.), die Chunqiu 春秋-Periode (770–476 v. Chr.) und die Zhanguo 战国-Periode (475–221 v. Chr.) unterteilt.

AC 车

车有六等之数:车轸四尺，谓之一等；戈柲六尺有六寸，既建而迤，崇于轸四尺，谓之二等；人长八尺，崇于戈四尺，谓之三等；殳长寻有四尺，崇于人四尺，谓之四等；车戟常，崇于殳四尺，谓之五等；酋矛常有四尺，崇於戟四尺，谓之六等。车谓之六等之数。

凡察车之道，必自载于地者始也，是故察车自轮始。凡察车之道，欲其朴属而微至。不朴属，无以为完久也；不微至，无以为戚速也。轮已崇，则人不能登也；轮已庳，则于马终古登阤也。故兵车之轮六尺有六寸，田车之轮六尺有三寸，乘车之轮六尺有六寸。六尺有六寸之轮，轵崇三尺有三寸也，加轸与轐焉，四尺也。人长八尺，登下以为节。

MC 造车

车上的尺寸可分成六个等级:车厢底部的边框离地四尺，是第一个等级；戈和戈柄长六尺六寸，斜插在车上，比车底框高四尺，这是第二等；人高八尺，站在车上，比戈高四尺，这是第三等；殳[46]长一丈二尺，比人高四尺，这是第四等；车戟长一丈六尺，比殳高四尺，这是第五等；酋矛长二丈，比戟高四尺，这是第六等。与车相关的这些数字，构成了车的六个等级。

凡检验车子的规则，必然是从被承载的部件开始，由此检验车子由车轮开始。凡检验车子的规则，要求它要坚固，[47] 而且接触地面的面积要小。不坚固，使用就不能持久；轮子着地面积不小，就不会运转迅捷。轮子太高大，人就不能登车；车轮太小，马拉着就像一直在爬坡一样。因此战车的车轮高六尺六寸，田猎用车车轮高六尺三寸，乘坐用车车轮高也是六尺六寸。六尺六寸高的车轮，车轴高三尺三寸，加上轸和轐，[48] 共高四尺。人的个子一般高八尺，这样上车下车的时候正好可以中间有一个踏脚缓冲的地方。

---

46 殳，古代一种兵器，用于撞击、打击，是后世棍、棒型兵器的前身。

47 《周礼注疏》郑玄注云:"朴属，犹附着坚固貌也。"

48 轐，汉代又叫 "伏兔"，是用来托载车厢的。古代车的结构是这样的：车轴上有伏兔，伏兔托载着车轸，车轸上面是车厢。车轴离地3尺3寸，伏兔和车轸合起来厚 7寸，它们和车轴高度加起来，正好离地4尺。

Manufacture of Chariots*

Often, a product can only be finished through the cooperation of several artificers. This is above all true for carriages. A carriage holds six difference sizes: the rear railing is four Chi 尺 (78.6 cm) high. This is the first size. The lance, including its grip, is six Chi 尺, six Cun 寸 (129.7 cm) long. If one puts it on the carriage at a slant, it sticks out four Chi 尺 above the rear railing. This is the second size. A man is eight Chi 尺 (157.2 cm) tall. The lance is four Chi 尺 higher than the man. This is the third size. The thrust weapon *Shu* 殳 is one Xun 寻, four Chi 尺 (235.8 cm) long. It is four Chi 尺 longer than a man. This is the fourth size. The halberds for the chariot are one Chang 常 (315.2 cm) and therefore four Chi 尺 longer than the weapon *Shu* 殳. This is the fifth size. The spear *Youmao* 酉矛 is one Chang 常, four Chi 尺 (393 cm) long and therefore four Chi 尺 longer than a halberd. This is the sixth size. Therefore, it is said that for a chariot, six different sizes apply.

If one looks at the essence of a carriage, one has to begin with its load on the earth. Therefore, the investigation of a carriage starts with the wheels. A carriage is required, to be firm and its wheels round. If it is not firm, it will not last long. If the wheels are not round, one cannot drive speedily. If the wheels are too large, it is difficult to get onto the carriage. If the wheels are too small, the horse must pull as hard as if it was climbing a mountain slope. Therefore, the war chariots have wheels with a diameter of six Chi 尺, six Cun 寸 (129.7 cm), whereas the hunting carriages have wheels with a diameter of six Chi 尺, three Cun 寸 (123.8 cm) and the wheels of travel carriages have a diameter of six Chi 尺, six Cun 寸. For wheels with a diameter of six Chi 尺, six Cun 寸, the axle bearing is three Chi 尺, three Cun 寸 (64.8 cm) high. If the rear railing and axle bearing are mounted on the carriage, one gets a height of four Chi 尺 (78.6 cm). It then becomes convenient for a man eight Chi 尺 tall to get on and off.

Wagenbau**

Ein Gegenstand kann oft erst durch die Zusammenarbeit mehrerer Handwerker fertiggestellt werden. Am meisten trifft das auf die Wagen zu. Am Wagen gibt es sechs verschiedene Differenzmaße: Die hintere Querstange des Wagens ist vier Chi 尺 (78,6 cm) hoch. Das ist das erste Maß. Die Lanze einschließlich der Griffstange ist sechs Chi 尺 sechs Cun 寸 (129,7 cm) lang. Wenn man sie

* See commentary: Manufacture of Carriages.

** Siehe Kommentar: Manufacture of Carriages.

schräg auf dem Wagen hinstellt, ragt sie vier Chi 尺 über der hinteren Querstange hervor. Das ist das zweite Maß. Ein Mann ist acht Chi 尺 (157,2 cm) groß. Die Lanze ist vier Chi 尺 größer als der Mann. Das ist das dritte Maß. Die Schlag- und Stichwaffe *Shu* 殳 ist ein Xun 寻 und vier Chi 尺 (235,8 cm) lang. Sie ist vier Chi 尺 größer als ein Mann. Das ist das vierte Maß. Die Hellebarden für den Wagen sind ein Chang 常 lang (315,2 cm) und somit vier Chi 尺 länger als die Waffe *Shu* 殳. Das ist das fünfte Maß. Der Speer *Youmao* 酉矛 ist ein Chang 常 und vier Chi 尺 (393 cm) lang und somit vier Chi 尺 länger als eine Hellebarde. Das ist das sechste Maß. Deshalb sagt man, dass für den Wagen sechs Differenzmaße gelten.

Wenn man das Wesen eines Wagens betrachtet, muss man mit der Last auf der Erde beginnen. Deshalb fängt die Untersuchung eines Wagens mit seinen Rädern an. Hinsichtlich der Forderungen an einen Wagen möchte man, dass er fest ist und die Räder rund sind. Ist er nicht fest, wird er nicht lange halten. Sind die Räder nicht rund, kann man nicht schnell fahren. Wenn die Räder zu groß sind, kann man den Wagen schlecht besteigen. Sind die Räder zu klein, dann muss sich das Pferd zu sehr anstrengen, als ob es einen Berg erklimmen soll. Deshalb haben die Kriegswagen Räder mit einem Durchmesser von sechs Chi 尺 und sechs Cun 寸 (129,7 cm), während die Räder von Jagdwagen einen Durchmesser von sechs Chi 尺 und drei Cun 寸 (123,8 cm) und die Räder von Reisewagen einen Durchmesser von sechs Chi 尺 und sechs Cun 寸 haben. Bei Rädern mit einem Durchmesser von sechs Chi 尺 und sechs Cun 寸 ist das Achslager drei Chi 尺 und drei Cun 寸 (64,8 cm) hoch. Bringt man auf dem Wagen noch die hintere Querstange und das Achslager an, so ergibt sich eine Höhe von vier Chi 尺 (78,6 cm). Bei einer Größe eines Mannes von acht Chi 尺 ist das Auf- und Absteigen dann bequem.

1AC. 轮人

轮人为轮，斩三材必以其时。三材既具，巧者和之。毂也者，以为利转也。辐也者，以为直指也。牙也者，以为固抱也。轮敝，三材不失职，谓之完。望而眡其轮，欲其幎尔而不[49]迤也。进而眡之，欲其微至也。无所取之，取诸圜也。望其辐，欲其掣尔而纤也。进而眡之，欲其肉称也。无所取之，取诸易直也。望其毂，欲其眼也，进而眡之，欲其帱之廉也。无所取之，取诸急也。眡其绠，欲其蚤之正也，察其菑蚤不齵，则轮虽敝不匡。

49 “不迤” 原作 “下迤”。据阮元《十三经注疏》，“下” 为 “不” 之误，故改为“不迤”。

凡斩毂之道，必矩其阴阳。阳也者，稹理而坚；阴也者，疏理而柔。是故以火养其阴，而齐诸其阳，则毂虽敝不蔽。毂小而长则柞，大而短则挚。是故六分其轮崇，以其一为之牙围；参分其牙围，而漆其二。椁其漆内而中诎之，以为之毂长。以其长为之围，以其围之防捎其薮。五分其毂之长，去一以为贤，去三以为轵。容毂必直，陈篆必正，施胶必厚，施筋必数，帱必负干。既摩，革色青白，谓之毂之善。参分其毂长，二在外，一在内，以置其辐。凡辐，量其凿深以为辐广。辐广而凿浅，则是以大扤，虽有良工，莫之能固。凿深而辐小，则是固有余，而强不足也，故竑其辐广以为之弱，则虽有重任，毂不折。参分其辐之长而杀其一，则虽有深泥，亦弗之溓也。参分其股围，去一以为骹围。揉辐必齐，平沈必均，直以指牙。牙得，则无槷而固；不得，则有槷必足见也。六尺有六寸之轮，绠参分寸之二，谓之轮之固。凡为轮，行泽者欲杼，行山者欲侔。杼以行泽，则是刀以割涂也，是故涂不附。侔以行山，则是搏以行石也，是故轮虽敝，不甐于凿。凡揉牙，外不廉而内不挫，旁不肿，谓之用火之善。是故规之以眡其圜也，萬之以眡其匡也，县之以眡其幅之直也，水之以眡其平沈之均也。量其薮以黍，以眡其同也，权之以眡其轻重之侔也。故可规、可萬、可水、可县、可量、可权也，谓之国工。

轮人为盖，达常围三寸，桯围倍之，六寸。信其桯围以为部广，部广六寸。部长二尺，桯长倍之，四尺者二。十分寸之一，谓之枚。部尊一枚，弓凿广四枚，凿上二枚，凿下四枚，凿深二寸有半，下直二枚，凿端一枚。弓长六尺谓之庇轵，五尺谓之庇轮，四尺谓之庇轸。参分弓长而揉其一，参分其股围，去一以为蚤围。参分弓长，以其一为之尊，上欲尊而宇欲卑，上尊而宇卑，则吐水疾而霤远。盖已崇，则难为门也；盖已卑，是蔽目也。是故盖崇十尺。良盖弗冒弗纮，殷亩而驰，不队，谓之国工。

1MC. 轮人

轮人制作车轮，伐取三材[50]必须按一定的时令进行。得到三材以后，能工巧匠对其进行巧妙地加工。毂的功能是使车轮旋转便利，辐的功能是在车轮中笔直地连接车毂与轮圈，牙的功能是使车轮外圈坚固地环抱在一起。车轮虽然已经使用得很破旧，但三材的功能仍然没有消失，这就叫完美。在一旁观测检视车轮，希望车轮各部位均匀密致，辐条挺直不扭。进一步近前检视，希望看到车轮与地面接触处尽可能地小。要做到这一

---

50 三材，指制作车轮的毂、辐、牙所用的木材。"毂" 是车轮中心的轴套部分，它内承车轴，外有榫眼， 以插辐条。"辐" 是车轮中连接毂和轮圈的直木条，轮圈的外部是"牙"，也叫做"輞"。

点，没有别的办法，就是要车轮尽可能地圆。在一旁观察车轮的辐条，希望它像人的手臂一样漂亮，[51] 但又下粗上细。进前细观，希望它各处匀称致密。[52] 没有别的要求，就是要求它要挺直。在一旁观看车毂，希望它大形整齐光洁。进前细观，希望幔毂的革均匀平整，包裹严密，没有起皱成棱的地方。没有别的指标要求，就是看它包裹得是否严紧。观察车轮的轮绠，希望辐条插入轮牙处坚固安稳，查看辐条两端插入轮牙和轮毂处的坚固情况，[53] 如果插入紧密，连接坚固，车轮虽然用旧也不会松动变形。

伐取用于制毂的木材，必须在伐树之前，先把木材的阴面和阳面标记下来。阳面部分纹理细密，木质坚实；阴面部分纹理粗疏，木质柔软，所以要用火烘烤阴面，使其坚实，与阳面性质一致，这样车毂即使用旧了，也不会变形。毂如果上下尺度较小而轴向尺度大，毂处辐条间距就会太窄，辐条插入毂的部分就不牢固；如果毂大而短，那么毂末浅短，亦不坚牢。因此车轮各部位要按一定的比例制作。六尺六寸高的车轮，将其分成六份，那么轮圈的尺度是其六分之一，即一尺一寸；把轮圈按其高度分成三份，内侧的三分之二涂上漆。测量着漆处的外径，取其一半的长度作为毂长。使毂的周长与其长度相等，用毂的周长的三分之一[54]作为安装辐条处的毂孔，毂长的五分之四作为毂孔内侧孔套（贤）的周长，毂长的五分之二作为毂孔外侧孔套（轵）的周长。[55] 制作毂一定要使毂孔匀直，使毂孔车轴密合；毂上的纹饰一定要周正，敷胶要厚实，缠筋要密实，毂上幔的皮革要紧贴毂体。敷完革后要摩平，摩完后革色青白，是毂的质量高的标志。把毂长分作三份，从外侧起量至三分之二的地方，是安装辐条的位置。辐条安装在榫眼中，榫眼的深度与其宽度应该一样。榫眼如果宽度大而深度浅，辐条安装上去容易动摇，这种情况下，即使技术再高超的工匠，也没办法使其牢固；榫眼深度深但宽度小，就会导致虽然安装的很牢固，但辐条的强度不够。因此安装辐条的榫眼的深度与其宽度要相等，[56] 这样即使车有重载，毂也不会毁坏。把辐条长度分成三份，其三分之一是安装在轮围和轮毂中的，这样即使车子在很深的烂泥中行驶，也不会被粘住。辐条要上细下粗，靠近轮围处的周长应是靠近轮毂处周长的三分之二，[57] 在制作辐条时，要使其规格一致，在水上的浮起程度也要一

51 掔，音 xiāo，本意是人的胳膊长得既长又好的样子。

52 肉，此处指辐条的各处；称，匀称之意。

53 "察其菑蚤不齵"，"菑"，此处指辐条插入轮毂的部分；"蚤"，此处指辐条插入轮围的部分；"齵"，本 意指齵齿，表示牙齿松动。故此句用 "不齵" 指辐条插入轮毂和轮围的坚固状况。

54 "防"，三分之一；"薮"，指车辐条会聚，此处引申为车毂的孔心。

55 贤，指轮毂孔内侧大口；轵，指轮毂孔外侧小口。不管是贤还是轵，都要在孔内衬垫金属套，以增加耐 磨性。金属套厚一般为 1 寸，故此处贤、轵的周长是包括了金属套在内的。

56 "竑其辐广以为之弱"，此处"弱"即前所谓之"菑"，即安装辐条之榫眼。

57 "参分其股围，去一以为骹围，" 郑玄《考工记注》引郑众云："股谓近毂者也，骹谓近牙者也。"

致，要垂直地安装在轮牙上。辐条与轮牙上的榫眼尺寸相得，不加楔子也很牢固；尺寸不相得，即使加了楔子，楔子也会显露出来。六尺六寸大小的轮子，在凿轮牙安装辐条时，凿孔要向外侵三分之二寸，这样轮子就坚固了。大凡是制作车轮，如果要行驶于泽地，轮子顶部要削尖；如果要行驶于山地，轮子顶部要平，使轮牙厚度上下相等。尖轮行驶于泽地，就像是用刀子切泥，泥不会粘附在轮子上；平轮在山地行驶，宽厚的车轮滚压在山石上，轮子即使破旧了，安装辐条的凿眼也不会变形而使辐条松动。凡是用火烘烤来揉制轮牙，做到外侧纹理不拉断，内侧纹理不锉折，旁边不凸起，就是善于用火的表现。检验车轮时，要用圆规来判断其是否正圆；将其安装在旋轮上旋转，观察其上下变化以判断其毂孔是否在正中心；将其悬垂以观察其辐条是否对直；让其浮在水上以观察其沉浮是否平均；用黍米测量两毂之孔容积是否相同；称量轮子的重量以判断两轮轻重是否一致。当车轮用这六种方法测量都能符合要求时，其制作者可谓是国之名匠了。

1. The Wheelmaker*

The wheelmaker manufactures the wheels. When he fells the three different kinds of wood for hub, spokes and rim, he has to consider the right season. In order to process the three kinds of wood, one has to use the right technology. The hub must enable the easy rotation of the wheel. The spokes must be straight. The rim should be held tightly together by its parts. If the wheel is worn, the three parts must still work. Then the wheel is perfect. If one looks at a wheel from afar, one must see whether it touches the ground evenly.[58] On closer view, one must see that the contact area of the wheel with the ground is small. In each case, one must demand that the wheel is round. If one looks at the spokes from afar, one must see that they taper like an arm. On closer view, they should be smooth and even. In each case, they must be straight. If one looks at the hub from afar, one must see that the leather cover has no edges. In each case, the hub must be covered tightly. If one looks at the rim from afar, one must see that the studs of the spokes in the rim are even. If one observes

---

* See commentary: Manufacture of Carriages.

58 *If one looks at a wheel from afar one must watch whether it touches the earth evenly* – This sentence is understood by different researchers in different ways. J. Needham meant that the wheels should look even, like a hanging curtain (*Science and Civilisation in China*, Vol. 4, p. 2 (1965), 75). Dutch scientist A.F.P. Hulsewé thought that if one looks at two rotating wheels mounted on an axle one can observe whether the Moiré pattern is even (*Mulun xingshi he zuoyong de yanbian* 木轮形式和作用的演变, Zhongguo keji shi tansuo 中国科技史探索 (1986), Shanghai guji chubanshe 上海古籍出版社, 453–484).

that the studs of the spokes in the hub are even, then even a worn wheel will not deform.

The essence of felling the wood for the hub is that one must mark the Yin 阴 and the Yang 阳 side of the wood.[59] On the Yang 阳 side, the grain of the wood is dense and firm, whereas the grain on the Yin 阴 side is wider and the wood softer. Therefore, the Yin 阴 side of the wood is roasted over a fire, so that the grain equals that on the Yang 阳 side. Even if the hub is worn, it will not warp. If the hub has a small diameter and is long, the distances between the spokes are too narrow. If the hub has a large diameter but is short, then the spokes are not seated tightly. Therefore, one sixth of the wheel's height yields the height of the rim. Two thirds of the rim's height is varnished from the inside, and the rest not. The diameter of the varnished wheel part is measured, then half of it is chosen as the length of the hub. The circumference of the hub equals its length. The circumference of the axle hole amounts to one third of the hub's circumference. Four fifths of the hub's length constitute the circumference of the axle hole on the inner side, and two fifths of the hub length constitute the circumference of the axle hole on the outer side.[60] The hub must be straight and the decorations on the outer side must be meticulous. The glue to join the studs must be applied thickly. The ribbons must be tied tightly. The leather cover must be glued firmly on the hub. If the leather is ground with a stone, it gets a pallid colour. Then one can speak of a good hub. One obtains the position of the spokes on the hub when the hub is divided into three parts: two parts outside, one part inside. In this way, the position of the spoke in the hub is determined. For the spokes, the depth of the stud hole in the hub equals the width of the spokes. If the depth of the stud hole is smaller than the width of the spoke, it results in heavy deformations of the wheel. Even the best workmanship does not get a stable wheel. If the stud hole is deep but the stud diameter is small, the joint is safe but not sufficiently firm, so that it can easily break. Therefore, one must measure to make sure that the spoke width equals the stud hole depth. Then the hub will not break, even with a heavy load on the carriage. At one third of its length, the spoke tapers towards the rim. If the carriage goes through deep mud, the wheel will not stick. The circumference of

59 *The Yin 阴 and the Yang 阳 side of the wood* – The side of a tree directed towards the sun is called the Yang 阳 side of the wood. The opposite side is correspondingly called the Yin 阴 side of the wood. The density of the wood's grain depends therefore on the influence of sun radiation.

60 *Four fifths of the hub's length constitute the circumference of the axle hole on the inner side, and two fifths of the hub length constitute the circumference of the axle hole on the outer side* – Although archaeological excavations of chariots have confirmed that the axle hole in the hub has a conical shape, the values given in the "Kao Gong Ji" are exaggerated.

the spokes on the rim amounts to two thirds of the spoke circumference on the hub. When processing the spokes, one must see to it that they are even. If one lays them in water, they must float at the same height. The spokes project into the rim. The stud hole in the rim and the stud must fit together. Even without a wooden wedge, they yield a firm joint. If stud hole and stud do not fit together, one must use wedges. For a wheel with a diameter of six Chi 尺, six Cun 寸 (129.7 cm), the inside inclination of the spokes in the direction of the hub amounts to 2/3 Cun 寸 (1.3 cm). In this way, one gets a stable wheel.[61]

Regarding wheels for driving through marshes, one needs a step-shaped cross section of the rim. On the other hand, for driving through mountains, one needs a rectangular cross section of the rim. The step-shaped cross section for driving through marshes cuts the mud like a knife so that one will not get stuck. The rectangular cross section for driving through mountains delivers a thick rim. Even if the wheel is worn, the spokes will not fall out of the stud holes. Generally, the rim is processed using a flame. In this way, the outside will not tear through tension, while the inside is not affected by the fire. The side planes will not swell through dry cracks. Those are the merits of treatment by a flame. Then the roundness of the wheel is tested with a circle gauge. With a planarity gauge, the correctness of the wheel is tested.[62] The wheel is suspended in order to test whether the spokes at the top and the bottom lead to each other. The wheel is laid in water to see whether it rises evenly from the water. The holes in the hub of the wheel are filled with millet seeds to test whether the volumes of the holes are even. On a balance, two wheels are weighed to see whether they are equally heavy. If one uses the circle and the planarity gauge, lays the wheels in water, suspends the wheels, measures the volume of the hub holes with millet seeds and weighs the wheels, and if the wheels pass the tests, one can call them excellent state products.

The carriage carpenter manufactures the carriage umbrella. On the top, the pole has a circumference of three Cun 寸 (5.9 cm). The lower pole has a double circumference that is six Cun 寸 (11.8 cm). The circumference of the lower pole provides the diameter of the hub for the spokes of the carriage umbrella, which amounts to six Cun 寸. The length of the upper pole, including the hub for the spokes, amounts to two Chi 尺 (39.3 cm). The lower pole consists of two sections, where each section is, with four Chi 尺 (78.6 cm), twice as long as the

---

61 *In this way one gets a stable wheel* – The slight inclination of the spokes to the inner sides gives the wheel more stability when the carriage takes a corner. The invention of dishing wheels in Europe is observed since the 17th century.

62 *With a planarity gauge the correctness of the wheel is tested* – The planarity gauge was laid on the wheel rim to test whether it laid in a plane.

upper pole, yielding a total length of eight Chi 尺 (157.2 cm). One tenth of a Cun 寸 (0.2 cm) is called one Mei 枚. The height of the cap on the hub for the spokes is one Mei 枚. The height of the rectangular holes in the hub where the spokes go is four Mei 枚 (0.8 cm). On top, the hole measures two Mei 枚, and on the bottom, four Mei 枚; the depth of the hole amounts to two-and-a-half Cun 寸 (4.9 cm). The holes taper so that their height at the end is two Mei 枚 (0.4 cm) and their width one Mei 枚. The spokes, with a length of six Chi 尺 (117.9 cm), are called "*bizhi*" 庇軹 (cover of the axle); with five Chi 尺 (98.3 cm) – "*bilun*" 庇輪 (cover of the wheels); and with four Chi 尺 (78.6 cm) – "*bizhen*" 庇軫 (cover of the carriage body). The spokes are bent at one third of their length from the hub. Two thirds of the spokes' circumference provide the circumference of the umbrella. The height difference between the hub and the ends of the spokes is one third of a spoke's length. The hub should be high and the ends of the spokes should be low, so that rainwater drains off quickly and flows away. If the carriage umbrella is too high, it is difficult to get through a gate; if it is too low, it obstructs the view. Therefore, for the height of the carriage umbrella ten Chi 尺 (196.5 cm) are laid down. A good carriage umbrella has neither curtains nor tassels on the spoke ends, so that it cannot fall down during the drive. Then it can be called an excellent state product.

1. Der Radmacher*

Der Radmacher fertigt die Räder an. Wenn er die drei verschiedenen Holzsorten für Radnabe, Speichen und Radkranz schlägt, muss er die richtige Jahreszeit beachten. Um die drei Holzsorten zu bearbeiten, muss man die richtige Technik anwenden. Die Radnabe muss eine leichte Drehung des Rades ermöglichen. Die Speichen müssen gerade sein. Der Radkranz muss aus seinen Einzelteilen fest zusammengefügt sein. Selbst wenn das Rad abgenutzt ist, müssen die drei Teile immer noch funktionieren. Dann ist ein Rad vollkommen. Betrachtet man ein Rad von Weitem, muss man darauf achten, ob es gleichmäßig die Erde berührt.[63] Von Nahem betrachtet, muss man darauf achten, dass die

* Siehe Kommentar: Manufacture of Carriages.

63 *Betrachtet man ein Rad von Weitem, muss man darauf achten, ob es gleichmäßig die Erde berührt* – Dieser Satz wird von verschiedenen Forschern unterschiedlich verstanden. J. Needham meinte, dass das Rad, wenn es aufgehängt wird, gleichmäßig wie ein hängender Vorhang aussehen muss. (*Science and Civilisation in China*, Vol. 4, p. 2 (1965), 75). Der niederländische Wissenschaftler Hulsewé meinte, dass, wenn man auf zwei sich drehende Räder schaut, die auf einer Achse montiert sind, man beobachten kann, ob das Moiré-Muster gleichmäßig ist (Mulun xingshi he zuoyong de yanbian 木轮形式和作用

Berührfläche des Rades auf der Erde klein ist. In jedem Fall muss man fordern, dass das Rad rund ist. Betrachtet man die Speichen von Weitem, muss man darauf achten, dass sie sich wie ein Arm verjüngen. Von Nahem betrachtet, sollen sie glatt und gleichmäßig sein. In jedem Fall müssen sie gerade sein. Betrachtet man die Radnabe von Weitem, muss man darauf achten, dass sie glatt und ebenmäßig ist. Von Nahem betrachtet, muss man darauf achten, dass das Abdeckleder auf der Nabe keine Kanten hat. In jedem Fall muss sie fest abgedeckt sein. Betrachtet man den Radkranz von Weitem, muss man darauf achten, dass die Zapfen der Speichen im Reifen gleichmäßig sind. Wenn man prüft, ob die Zapfen der Speichen in der Nabe gleichmäßig sind, dann wird sich auch ein abgenutztes Rad nicht verformen.

Das Wesen beim Schlagen des Holzes für die Radnabe besteht darin, dass man die Yin 阴- und die Yang 阳-Seite des Holzes markieren muss.[64] Auf der Yang 阳-Seite ist die Maserung des Holzes dicht, und es ist fest, während die Maserung auf der Yin 阴-Seite weiter und das Holz weicher ist. Deshalb röstet man die Yin 阴-Seite des Holzes über einem Feuer, so dass die Maserung der auf der Yang 阳-Seite gleichkommt. Selbst wenn die Nabe abgenutzt ist, wird sie sich nicht verziehen. Wenn die Nabe einen kleinen Durchmesser hat und lang ist, sind die Abstände zwischen den Speichen zu eng. Hat die Nabe einen großen Durchmesser, ist aber kurz, dann sitzen die Speichen nicht fest. Deshalb ergibt ein Sechstel der Höhe des Rades die Höhe des Radkranzes. Zwei Drittel der Radkranzhöhe von innen werden lackiert. Man misst den Durchmesser des lackierten Radteils und wählt die Hälfte davon als Länge der Nabe. Der Umfang der Nabe ist gleich ihrer Länge. Der Umfang der Achsbohrung beträgt ein Drittel des Nabenumfangs. Vier Fünftel der Nabenlänge bilden den Umfang der Achsbohrung auf der Innenseite, zwei Fünftel der Nabenlänge bilden den Umfang der Achsbohrung auf der Außenseite.[65] Die Nabe muss gerade sein, und die Verzierungen auf der Außenseite müssen akkurat sein. Der Leim zum Verbinden der Zapfen muss dick aufgetragen werden. Die Bänder müssen dicht gewickelt werden. Das Abdeckleder muss fest auf die

的演变, Zhongguo keji shi tansuo 中国科技史探索 (1986), Shanghai guji chubanshe 上海古籍出版社, 453–484).

64 *Dass man die Yin 阴- und die Yang 阳-Seite des Holzes markieren muss* – Die der Sonne zugewandte Seite eines Baums wird Yang 阳-Seite genannt und die gegenüberliegende Seite entsprechend Yin 阴-Seite. Die Dichte der Maserung des Holzes hängt auch vom Einfluss der Sonnenstrahlung ab.

65 *Vier Fünftel der Nabenlänge bilden den Umfang der Achsbohrung auf der Innenseite, zwei Fünftel der Nabenlänge bilden den Umfang der Achsbohrung auf der Außenseite* – Obwohl archäologische Ausgrabungen von Streitwagen bestätigt haben, dass die Nabenbohrung leicht konisch ist, sind die in "Kao Gong Ji" angegebenen Werte übertrieben.

Nabe geklebt werden. Wenn das Leder mit einem Stein geschliffen wird, nimmt es eine bleiche Farbe an. Dann kann man von einer guten Nabe sprechen. Die Nabenlänge wird in drei Teile unterteilt, zwei Teile außen und ein Teil innen. So wird die Lage der Speiche in der Nabe bestimmt. Für die Speichen bemisst man die Zapfenlochtiefe in der Nabe gleich der Breite der Speiche. Wenn die Zapfenlochtiefe kleiner als die Breite der Speiche ist, kommt es zu starken Verformungen des Rades. Selbst bei bester Verarbeitung erhält man dann kein stabiles Rad. Wenn die Zapfenbohrung tief, aber der Zapfendurchmesser klein ist, ist die Verbindung zwar sicher, aber nicht genügend fest, denn sie kann leicht brechen. Deshalb muss man messen, dass die Speichenbreite gleich der Zapfenlochtiefe ist. Dann wird die Nabe selbst bei einer schweren Last auf dem Wagen nicht brechen. Die Speiche wird auf einem Drittel ihrer Länge zum Radkranz hin schlanker gemacht. Wenn der Wagen durch tiefen Schlamm fährt, wird das Rad nicht festkleben. Der Umfang der Speichen am Radkranz beträgt zwei Drittel des Speichenumfangs an der Nabe. Bei der Bearbeitung der Speichen muss man darauf achten, dass sie gleichmäßig sind. Wenn man sie in Wasser legt, müssen sie gleichmäßig hoch schwimmen. Die Speichen ragen gerade in den Radkranz. Die Zapfenbohrung im Radkranz und der Zapfen müssen zueinander passen. Auch ohne einen Holzkeil ergeben sie eine feste Verbindung. Wenn Zapfenbohrung und Zapfen nicht passen, muss man Keile verwenden. Bei einem Rad mit einem Durchmesser von sechs Chi 尺 sechs Cun 寸 (129,7 cm) beträgt die Neigung der Speichen nach innen in Richtung der Nabe 2/3 Cun 寸 (1,3 cm). So erhält man ein stabiles Rad.[66]

Hinsichtlich der Räder braucht man beim Fahren in Sümpfen einen gestuften Querschnitt des Radkranzes. Demgegenüber braucht man beim Fahren durch Gebirge einen rechteckigen Querschnitt des Radkranzes. Der gestufte Querschnitt beim Fahren in Sümpfen zerschneidet den Schlamm wie ein Messer, so dass man nicht in ihm steckenbleibt. Der rechteckige Querschnitt beim Fahren im Gebirge liefert einen dicken Radkranz. Auch wenn das Rad abgenutzt ist, fallen die Speichen nicht aus den Zapfenlöchern. Allgemein wird der Radkranz mit einer Flamme behandelt. Dann wird die Außenseite durch Zug nicht reißen und die Innenseite durch das Feuer nicht beeinträchtigt. Die Seitenflächen quellen nicht durch Trockenrisse auf. Das sind die Vorzüge der Behandlung mit einer Flamme. Deshalb wird die Rundheit des Rades mit einer Kreislehre geprüft. Mit einer Ebenheitslehre wird die Korrektheit des

66 *So erhält man ein stabiles Rad* – Die leichte Neigung der Speichen nach innen gibt dem Rad mehr Stabilität, wenn man Kurven fährt. Diese Erfindung ist in Europa erst ab dem 17. Jahrhundert verbreitet.

Rades geprüft.[67] Man hängt das Rad auf, um zu prüfen, ob die Speichen oben und unten gerade aufeinander zu laufen. Man legt das Rad ins Wasser, um zu sehen, ob es gleichmäßig aus dem Wasser ragt. Man füllt die Bohrungen in den Naben der Räder mit Hirsekörnern, um zu prüfen, ob die Volumina der Bohrungen gleich sind. Auf einer Waage wiegt man zwei Räder, um zu sehen, ob sie gleich schwer sind. Wenn man die Rundheits- und die Ebenheitslehre benutzt, die Räder in Wasser legt, die Räder aufhängt, das Volumen der Nabenbohrungen mit Hirsekörnern misst und die Räder wiegt und die Räder die Prüfungen bestehen, kann man sie als vorzügliches Landesprodukt bezeichnen.

Der Wagenbauer fertigt den Wagenschirm an. Die Stange hat oben einen Umfang von drei Cun 寸 (5,9 cm), die untere Stange hat einen doppelt so großen Umfang, der sechs Cun 寸 (11,8 cm) beträgt. Die Länge der unteren Stange liefert den Durchmesser der Nabe für die Speichen des Wagenschirms, der somit sechs Cun 寸 beträgt. Die Länge der oberen Stange zusammen mit der Nabe für die Speichen beträgt zwei Chi 尺 (39,3 cm). Die untere Stange besteht aus zwei Abschnitten, wobei jeder Abschnitt mit vier Chi 尺 (78,6 cm) doppelt so lang wie die obere Stange ist, was eine Gesamtlänge von acht Chi 尺 (157,2 cm) ergibt. Ein Zehntel Cun 寸 (0,2 cm) heißt ein Mei 枚. Die Höhe der Kappe auf der Nabe für die Speichen beträgt ein Mei 枚. Die Höhe der viereckigen [trapezförmigen] Löcher in der Nabe zur Aufnahme der Speichen beträgt vier Mei 枚 (0,8 cm). Oben ist das Loch zwei Mei 枚 und unten vier Mei 枚 groß, die Tiefe des Loches beträgt zwei ein halb Cun 寸 (4,9 cm). Die Löcher verjüngen sich, so dass seine Höhe am Ende zwei Mei 枚 (0,4 cm) und seine Breite ein Mei 枚 beträgt. Die Speichen mit einer Länge von sechs Chi 尺 (117,9 cm) heißen „*bizhi*" 庛軹 (Abdeckung der Achse), mit fünf Chi 尺 (98,3 cm) – „*bilun*" 庛輪 (Abdeckung der Räder) und mit vier Chi 尺 (78,6 cm) – „*bizhen*" 庛軫 (Abdeckung des Wagenkastens). Die Speichen werden bei einem Drittel ihrer Länge von der Nabe aus gebogen. Zwei Drittel des Umfangs der Speichen bilden den Umfang des Schirms. Die Höhendifferenz zwischen der Nabe und den Enden der Speichen beträgt ein Drittel der Speichenlänge. Die Nabe soll hoch und die Enden der Speichen sollen niedrig sein, damit Regenwasser schnell abläuft und weit wegspritzt. Wenn der Wagenschirm zu hoch ist, gelangt man schlecht durch ein Tor, ist er zu niedrig, versperrt er die Sicht. Deshalb ist die Höhe des Wagenschirms mit zehn Chi 尺 (196,5 cm) festgelegt. Ein guter Wagenschirm hat weder Vorhänge noch Troddeln an den Speichenenden, so dass der Wagenschirm beim Fahren nicht umfallen kann. Dann kann man ihn als ein vorzügliches Landesprodukt bezeichnen.

---

67 *Wird die Korrektheit des Rades geprüft* – Eine Ebenheitslehre wurde auf den Radreifen gelegt, um zu prüfen, ob er sich in einer Ebene befindet.

2AC. 舆人

舆人为车，轮崇、车广、衡长，参如一，谓之参称。参分车广，去一以为隧。参分其隧，一在前，二在后，以揉其式。以其广之半为之式崇，以其隧之半为之较崇。六分其广，以一为之轸围。参分轸围，去一以为式围。参分式围，去一以为较围。参分较围，去一以为轵围。参分轵围，去一以为轛围。圜者中规，方者中矩，立者中县，衡者中水，直者如生焉，继者如附焉。凡居材，大与小无并。大倚小则摧，引之则绝。栈车欲弇，饰车欲侈。

2MC. 舆人

舆人制作车子，车轮的高、车厢的宽和车辕前端的横木，三者一样长，叫做三称。车厢的宽度减去三分之一，得到 "隧"（车厢）长。把车厢长分成三份，三分之一在前，三分之二在后，将 "式"（车厢栏杆）揉曲确定到这个位置。以车身宽度的一半作为 "式"（车厢栏杆）的高度，以车厢长的一半作为 "较"[68] 的高。以车厢宽度的六分之一作为 "轸"（车厢四周的横木）的周长。以 "轸" 周长的三分之二作为 "式" 的周长，"式" 周长的三分之二是 "较" 的周长，"较" 周长的三分之二是 "轵"（车厢两侧横直交结的栏木）的周长，"轵" 的周长的三分之二是 "轛"（车厢栏杆下面横直交结的栏木）的周长。圆的部件要合乎圆规，方的要合乎矩尺，竖立的要与悬锤一直，横放的要处于水平，竖立的就像树木从地上长出来一样，相互联结的就像树木的枝杈长在树上一样。凡是处理造车用材，大材和小材不能直接装配。如果二者不相称，用小材支撑大材，小材会被摧毁；让二者衔接牵连，小材会拉断。一般的士子所乘坐的 "栈车"，不用皮革缠绳扎系，应使其狭窄内敛；大夫等所乘坐的 "饰车"，用皮革缠绳扎系装饰，应将其造得宽敞考究。

2. The Carriage Carpenter*

The carriage carpenter manufactures the carriage body. The height of the wheels, the width of the carriage body and the transverse axis of the pole – all three are equal. Therefore, they are called the three equals. Two thirds of the carriage body's width provide its length. If one divides the length of the

68 较，依据贾公彦的疏解，是指车厢两旁竖立的短柱，短柱上又伸出的一节。上面用横栏连接。

* See commentary: Manufacture of Carriages.

carriage body into three parts, then one third lies before and two thirds behind the axle. The front transverse rod is bent to the sides. The height of the front transverse rod is half the carriage body's width. Half the carriage body's length provides the height of the side holds. One sixth of the carriage body's width provides the circumference of the rear railing. The circumference of the front transverse rod is two thirds of the circumference of the side holds. Two thirds of the circumference of the side holds provide the circumference of the side railing. Two thirds of the circumference of the side railing provide the circumference of the front railing.

Roundness is obtained with a pair of compasses, a square with a right angle, a perpendicular with a hanging string, a horizontal with a water level and a straight line is obtained when a plant grows from the earth. These shapes are connected with each other like branches with a trunk. If, when processing the materials for building a carriage, large and small do not fit together, one cannot assemble it. If a small part supports a large one, it can break. If a small part draws a large one, it can tear. The carriages of officials, made from bamboo and other materials, should be somewhat smaller; the carriages of dignitaries, covered with leather, should be somewhat larger.

2. Der Wagenbauer*

Der Wagenbauer fertigt den Wagenkasten an. Die Höhe der Räder, die Breite des Wagenkastens und das Querholz der Deichsel sind alle drei gleich. Deshalb heißen sie die drei Gleichen. Zwei Drittel der Breite des Wagenkastens ergeben seine Länge. Wenn man die Länge des Wagenkastens in drei Teile teilt, so liegen ein Drittel vor und zwei Drittel hinter der Achse. Das Querholz über dem vorderen Ende des Wagenkastens wird zu den Seiten herumgebogen. Die Höhe des Querholzes beträgt die Hälfte der Breite des Wagenkastens. Die Hälfte der Länge des Wagenkastens ergibt die Höhe der seitlichen Haltegriffe. Ein Sechstel der Breite des Wagenkastens ergibt den Umfang der hinteren Querstange. Der Umfang der vorderen Querstange beträgt zwei Drittel des Umfangs der seitlichen Haltegriffe. Zwei Drittel des Umfangs der seitlichen Haltegriffe ergeben den Umfang der seitlichen Geländer. Zwei Drittel des Umfangs der seitlichen Geländer ergeben den Umfang des vorderen Geländers.

Die Rundheit erhält man mit einem Zirkel, das Quadrat mit einem rechten Winkel, die Senkrechte mit einer hängenden Schnur, die Horizontale mit einem Wasserspiegel und die Gerade ergibt sich, wenn eine Pflanze aus der

* Siehe Kommentar: Manufacture of Carriages.

Erde wächst. Diese Formen sind miteinander verbunden wie die Zweige mit dem Stamm. Wenn bei der Verarbeitung der Materialien für den Wagenbau Groß und Klein nicht zusammenpassen, kann man ihn nicht zusammenbauen. Wenn ein kleines Teil ein großes stützt, kann es brechen. Wenn ein kleines Teil ein großes zieht, kann es reißen. Die aus Bambus und anderen Materialien hergestellten Wagen der Beamten sollten etwas kleiner, die mit Leder verkleideten Wagen der Würdenträger können etwas größer sein.

3AC. 辀人

辀人为辀。辀有三度，轴有三理。国马之辀，深四尺有七寸，田马之辀深四尺，驽马之辀深三尺有三寸。轴有三理:一者，以为媺也；二者，以为久也；三者，以为利也。軓前十尺，而策半之。凡任木，任正者，十分其辀之长，以其一为之围。衡任者，五分其长，以其一为之围。小于度，谓之无任。五分其轸间，以其一为之轴围。十分其辀之长，以其一为之当兔之围。参分其兔围，去一以为颈围。五分其颈围，去一以为踵围。

凡揉辀，欲其孙而无弧深。今夫大车之辕挚，其登又难，既克其登，其覆车也必易。此无故，唯辕直且无桡也。是故大车，平地既节轩挚之任，及其登阤，不伏其辕，必缢其牛。此无故，唯辕直且无桡也。故登阤者，倍任者也，犹能以登，及其下阤也，不援其邸，必緧其牛后。此无故，唯辕直且无桡也。是故辀欲颀典，辀深则折，浅则负。辀注则利准，利准则久，和则安。辀欲弧而无折，经而无绝，进则与马谋，退则与人谋，终日驰骋，左不楗。行数千里，马不契需；终岁御，衣衽不敝:此唯辀之和也。劝登马力，马力既竭，辀犹能一取焉。良辀环灂，自伏兔不至軓七寸，軓中有灂，谓之国辀。

轸之方也，以象地也；盖之圜也，以象天也；轮辐三十，以象日月也；盖弓二十有八，以象星也；龙旂九斿，以象大火也；鸟旟七斿；以象鹑火也；熊旗六斿，以象伐也；龟蛇四斿，以象营室也；弧旌枉矢，以象弧也。

3MC. 辀人

辀人负责做辀[69]（车的曲辕部分）。车辕的弯曲有三种弧度，车轴的质量有三个指标。优良的国马所驾之车，辀的曲高为四尺七寸；狩猎用马所驾

69 这里的“辀”车，指的是春秋战国时期的独辕车。这种车由一根连接在车轴和车厢上的独辕牵动，独　　辕前半部分向上弯曲，其顶端缚有横木，以供马牵拉。

之车，辀的曲高应为四尺；一般驽马所驾之车，辀的曲高为三尺三寸。高质量车轴要满足三个指标，第一是木材纹理均匀，没有结疤；第二是木质坚韧，可以持久，不会变形；第三是木质细密坚硬，与毂配合良好，使车轮旋转滑利。軓（车厢底前面的横木，是车厢最靠前的部件）距离辀端十尺，马鞭的长度是其一半，即五尺。凡是造车时要负重的木材，特别是直接承担车的重量的木材，其周长应该是辀的长度的十分之一。[70] 而驾车时搁在马颈上的横木，其周长是其长度的五分之一。[71] 小于这个尺度，它就不能胜任。以两轸（车厢底部四周的横木）间距离的五分之一作为车轴的周长，以辀长度的十分之一为当兔[72]的周长。以当兔周长的三分之二作为辀颈（辀的前端稍凹以用于持衡的部位）的周长，以辀颈的周长的五分之四作为辀踵（辀的后端用于连接后面车轸的部位）的周长。

凡是用火烤的方法使辀弯曲，希望木材的纹理自然弯曲而不过度。[73] 现在的牛车，辕是直的，其车前低后高，[74] 上坡很难。即使能上坡，也容易翻车。这没有别的原因，就是因为它的辕是直的，没有弯曲的缘故。因此对于牛车，平地上行驶时，可以前后轻重相宜，[75] 胜任负载，到了上坡时，如果不压抑车辕，一旦车辕翘起，就会勒住牛的脖子。这没有别的原因，就是因为它的辕是直的，没有弯曲的缘故。因此上坡时，虽然成倍费力，也还能登上去，一旦下坡，如果不拉住车尾，拴套在牛身上的革带必然会从后勒赶牛身，这没有别的原因，就是因为它的辕是直的，没有弯曲的缘故。因此车辀一定要坚韧，并有一定的弯曲度。如果弯曲度太大，就容易折断；如果弯曲度太小，车前重后轻，马就会像负重一样，容易疲累。辀的弯曲程度深浅合适，就会行走快捷平稳，快捷平稳就会持久，弯曲程度自然，车就安稳。辀的弧度要顺着木的纹理自然弯曲，没有曲折，木的纹理也没有断绝，车的进退都和马及御者之意相合，这样一天到晚驰驱不停，左侧的骖马也不会疲倦。[76] 即使驰行了几千里，马蹄也不会受伤；[77] 御者一年到头御车，衣服也不会破烂，这都是因为车辀做得好的缘故。驱

70 郑玄《考工记注》云："辀，軓前十尺与隧四尺四寸，凡丈四尺四寸。则任正之围，尺四寸五分寸之二。"即造车时承重材料的周长应为 1尺 4寸 4分。

71 郑玄《考工记注》云："衡任者，谓两軶之间也。兵车、乘车衡围一尺三寸五分寸之一。""軶"是套在 马颈上的人字形马具。

72 当兔，指垫在辀与轴之间的木块。其上面为纵向凹槽，以便承辀；下面为横向凹槽，以便含轴。

73 "欲其孙"指顺着木材的纹理使其弯曲。

74 郑玄注，"挚，輖也。"輖意为车重，即车辕重，车前重后轻之意。

75 轩挚，即轩轾，车舆前轻后重曰轩，前重后轻曰轾。"节轩挚之任"，是说能够调节前后轻重，胜任负 载。

76 古代战车一般用 4 皮马驾辕。中间的两匹称为服马，外侧的两匹称为骖马。因为御者在左，左侧骖马 对御者意图需做出更迅速的反应，因而比其余 3匹马更容易疲累。

77 《周礼注疏》引郑众曰："契读为'爰契我龟'之契，需读为'畏需'之需。谓不伤蹄，不需道里。"

使马儿用力，即使马的力气用尽了，车还能向前运行一段距离。好的车辀，上面的油漆如环。但从伏兔至近軓（车箱前面的档板）的 7 寸长度处没有漆。軓处漆纹完整的，就可称之为国家水平的优秀车辀了。

车厢是方形的，象征大地；车盖是圆形的，象征上天。车轮的辐条共三十根，象征日月；车盖的弓条有二十八根，象征天上的二十八宿。龙旂（旗）绘龙的图案，饰九斿，象征大火星；鸟旗绘有鸟隼图案，饰七斿，象征鹑火星；熊旗绘有熊的图案，饰六斿，象征伐星；龟旗绘有龟蛇图案，饰四斿，象征营室星；弧旗绘有弓箭图案，象征弧星。

3. The Pole Carpenter*

The pole carpenter manufactures the pole. For the pole, three different measures are relevant, and for the axle, three different quality criteria. The pole for excellent horses has a height of four Chi 尺, seven Cun 寸 (92.4 cm) above earth; for hunting horses, a height of four Chi 尺 (78.6 cm); and for ordinary horses, a height of three Chi 尺, three Cun 寸 (64.8 cm). For the axle, three criteria are valid: first, the grain of the wood must be even and it must not have any knots; second, the wood must be firm; and third, axle and hub must be joined tightly. The length of the pole from the step before the carriage body is ten Chi 尺 (196.5 cm); the length of the whip is one half of it. Generally, the wooden parts of the carriage constitute the load, which the parts under the carriage bear. Therefore, poles are needed with a circumference of one tenth of the pole length. The cross rod between the two horse collars has a circumference that is one fifth of its length. If a smaller size is chosen, then the parts cannot bear the load. One fifth of the length of the cross rod between two poles delivers its circumference. The circumference of the wooden block that fixes the pole is one tenth of the pole length. Two thirds of the circumference of this wooden block provide the circumference of the pole on the front end. Four fifths of the circumference of the pole's front end provide the circumference of the pole on the rear end.

The pole is bent over fire. One must follow the grain of the wood and must not bend it excessively. Today, the straight poles of large carriages are placed low, so that it is difficult to ascend a slope. And if one drives up, the carriage can easily fall over. For this, there is no other reason than that the pole is straight and not bent. Therefore, such large carriages are suited to transport evenly distributed loads on even terrain. But if one wants to take a slope and the driver does not press down the pole, it chokes the neck of the buffalo. This is caused

* See commentary: Astronomy.

by no other reason than that the pole is straight and not bent. Therefore, one can take the slope only if the animal doubles its force. If one drives down a slope, the driver must brake from behind, otherwise the carriage will push against the hind legs of the animal. For this, there is no other reason than that the pole is straight and not bent. Therefore the pole should be strong and bent in a suitable way. If the pole is bent too sharply, it can easily break. If the curvature is not sufficient, the carriage will roll over to the back. If the curvature of the pole on the front end equals the line of the constellation *Zhu* 注, one can drive fast. The back part of the pole is horizontal and durable. If straightness and curvature are adjusted to each other, the pole is stable. If the pole is bent in a suitable way, it will not break. There must be no cracks along the grain. Man and horse must cooperate without constraint. If one drives an entire day in this way, the left horse of the team[78] will not be tired. Even if one drives several thousand Li 里, the horse will not hurt its hooves and get nervous. Even if one drives a whole year, the clothing of the driver will not wear out. This is because the pole is bent in a suitable way! A good pole is conducive to developing the force of the horse. Even if the horse does not draw further, the carriage still rolls on quite a bit.[79] A good pole is lacquered with the patterns of jade rings. Seven Cun 寸 (13.8 cm) from the wooden block that the pole is fixed to, the rear part of the pole … [part is missing]. If this wooden block is lacquered as well, one can call it an excellent pole.

The rectangular shape of the carriage body symbolises earth, and the round carriage umbrella sky. The thirty spokes of the wheel symbolise the thirty days of a month, and the twenty-eight spokes of the carriage umbrella the twenty-eight lunar mansions. The dragon banner is decorated with nine ribbons, which symbolise the star Great Fire (*da huo* 大火). The falcon banner is decorated with seven ribbons, which symbolise the star Quail Fire (*chun huo* 鶉火). The bear banner is decorated with six ribbons, which symbolise the star Chopper (*fa* 伐). The snake banner is decorated with four ribbons, which symbolise the star Quarters House (*ying shi* 營室). The bow-and-arrow banner is decorated with the arrow *wang shi* 枉矢, which symbolises the star Arc (*hu* 弧).

78 *Left horse of the team* – The chariots had three horses in the team. The driver stood on the left side of the chariot and therefore observed the left horse most attentively.

79 *Even if the horse does not draw further, the carriage still rolls on quite a bit* – Here we have the earliest description of the phenomenon of inertness in a Chinese source.

3. Der Deichselbauer*

Der Deichselbauer fertigt die Deichsel an. Für die Deichsel sind drei verschiedene Maße und drei verschiedene Gütekriterien maßgeblich. Die Deichsel für ausgezeichnete Pferde hat eine Höhe über der Erde von vier Chi 尺 sieben Cun 寸 (92,4 cm), für Jagdpferde eine Höhe von vier Chi 尺 (78,6 cm) und für gewöhnliche Pferde eine Höhe von drei Chi 尺 drei Cun 寸 (64,8 cm). Für die Achse gelten drei Kriterien: Erstens muss die Maserung des Holzes gleichmäßig sein, und es darf keine Äste aufweisen, zweitens muss das Holz fest sein, und drittens müssen Achse und Nabe fest gefügt sein. Die Länge der Deichsel ab der Stufe vor dem Wagenkasten beträgt zehn Chi 尺 (196,5 cm), die Länge der Peitsche ist die Hälfte davon. Im Allgemeinen bewirken die Holzteile auf dem Wagen eine Last, die die Teile unter dem Wagen tragen. Hierfür braucht es Deichseln mit einem Umfang gleich einem Zehntel der Deichsellänge. Die Querstange zwischen den beiden Kummets hat einen Umfang, der ein Fünftel seiner Länge beträgt. Wählt man ein kleineres Maß, so können die Teile die Last nicht aufnehmen. Ein Fünftel des Abstands zwischen den beiden Querhölzern bildet den Umfang der Achse. Der Umfang der Holzblöcke zur Befestigung der Deichseln beträgt ein Zehntel der Deichsellänge. Zwei Drittel des Umfangs dieser Holzblöcke ergeben den Umfang der Deichsel am vorderen Ende. Vier Fünftel des Umfangs am vorderen Deichselende ergeben den Umfang der Deichsel an ihrem hinteren Ende. Die Deichsel wird über Feuer gebogen, dabei muss man der Maserung folgen und darf nicht übermäßig biegen. Heutzutage setzen die geraden Deichseln der großen Wagen tief an, so dass ein Anstieg schwierig zu nehmen ist. Und wenn man hochfährt, kann der Wagen leicht umstürzen. Hierfür gibt es keinen anderen Grund, als dass die Deichseln gerade und nicht gebogen sind. Darum sind solche großen Wagen geeignet, um mit gleichmäßig verteilter Last auf ebenem Gelände Lasten zu befördern. Aber wenn man einen Anstieg nehmen will und der Wagenführer die Deichsel vorn nicht herunterdrückt, schnürt sie dem Büffel den Hals ab. Hier wirkt kein anderer Grund, als dass die Deichsel gerade und nicht gebogen ist. Deshalb kommt man einen Anstieg nur hinauf, indem das Tier seine Kräfte verdoppelt. Und wenn man einen Hang hinunterfährt, muss der Wagenführer den Wagen von hinten bremsen, weil der Wagen sonst dem Tier gegen die Hinterbeine stößt. Hierfür gibt es keinen anderen Grund, als dass die Deichsel gerade und nicht gebogen ist. Deshalb sollte die Deichsel fest und geeignet gebogen sein. Ist die Deichsel zu stark gebogen, kann sie leicht brechen. Ist die Krümmung nicht ausreichend, wird der Wagen nach hinten kippen. Wenn

* Siehe Kommentar: Astronomy.

die Krümmung der Deichsel am vorderen Ende der Linie des Sternenbilds *Zhu* 注 gleicht, kann man schnell fahren. Der hintere Teil der Deichsel ist horizontal und dauerhaft. Wenn Geradheit und Krümmung aufeinander abgestimmt sind, ist die Deichsel stabil. Ist die Deichsel geeignet gekrümmt, wird sie nicht brechen. Entlang der Maserung dürfen keine Risse vorhanden sein. Mensch und Pferd müssen ohne Zwang zusammenarbeiten. Wenn man so den ganzen Tag fährt, wird das linke Pferd des Gespanns[80] keine Müdigkeit zeigen. Selbst wenn man mehrere tausend Li 里 fährt, wird sich das Pferd die Hufe nicht verletzen und ängstlich werden. Selbst wenn man das ganze Jahr fährt, wird die Kleidung des Wagenführers nicht verschleißen. Das liegt daran, dass die Deichsel geeignet gebogen ist! Eine gute Deichsel ist vorteilhaft, um die Kraft des Pferdes zu entfalten. Selbst wenn das Pferd nicht mehr zieht, rollt der Wagen noch ein Stück weiter.[81] Eine gute Deichsel ist mit den Mustern von Jaderingen lackiert. Sieben Cun 寸 (13,8 cm) vom Holzblock, an dem die Deichsel befestigt ist, ist der hintere Teil der Deichsel … [hier fehlt ein Teil] Wenn auch dieser Holzblock lackiert ist, dann kann man von einer vorzüglichen Deichsel sprechen.

Die rechteckige Form des Wagenkastens symbolisiert die Erde und der runde Wagenschirm den Himmel. Die dreißig Speichen des Rades symbolisieren die dreißig Tage eines Monats und die achtundzwanzig Speichen des Wagenschirms die achtundzwanzig Sternhäuser. Das Drachenbanner ist mit neun Bändern geschmückt, das den Stern Großes Feuer (*da huo* 大火) symbolisiert. Das Falkenbanner ist mit sieben Bändern geschmückt, das den Stern Wachtelfeuer (*chun huo* 鶉火) symbolisiert. Das Bärenbanner ist mit sechs Bändern geschmückt, das den Stern Holzfäller (*fa* 伐) symbolisiert. Das Schlangenbanner ist mit vier Bändern geschmückt, das den Stern Quartierhaus (*ying shi* 營室) symbolisiert. Das Pfeil-und-Bogen-Banner ist mit dem Pfeil *wang shi* 枉矢 geschmückt, das den Stern Bogen (*hu* 弧) symbolisiert.

4AC. 攻金

攻金之工，筑氏执下齐，冶氏执上齐，凫氏为声，栗氏为量，段氏为镈器，桃氏为刃。金有六齐:六分其金而锡居一，谓之钟鼎之齐；五分其金

80 *Linke Pferd des Gespanns* – Die Streitwagen hatten drei Pferde eingespannt. Der Wagenlenker stand auf der linken Seite des Wagens und beobachtete deshalb das linke Pferd am aufmerksamsten.

81 *Selbst wenn das Pferd nicht mehr zieht, rollt der Wagen noch ein Stück weiter* – Dies ist die früheste Beschreibung des Phänomens der Trägheit in einer chinesischen Quelle.

而锡居一，谓之斧斤之齐；四分其金而锡居一，谓之戈戟之齐；参分其金而锡居一，谓之大刃之齐；五分其金而锡居二，谓之削杀矢之齐；金锡半，谓之鉴燧之齐。

4MC. 攻金

负责制作金属器物的工匠有六类：筑氏负责下齐（齐，通 "剂"，指铜锡合金中锡的比例。下齐，指锡的比例较大的青铜合金），冶氏负责上齐（铜锡合金中锡的比例较小），凫氏制作乐器，栗氏制作量器，段氏制作农具， 桃氏制作兵器。青铜合金有六种比例：锡占六分之一的，是钟鼎之齐，适宜于制作钟、鼎等器物；锡占五分之一的，是斧斤之齐，适宜于制作斧斤等砍伐工具；锡占四分之一的，是戈戟之齐，适宜于制作戈、戟等兵器；锡占三分之一的，是大刃之齐，适宜于制作刀剑等带利刃的兵器；锡占五分之二的，是削杀矢之齐，适宜于制作小刀、镞矢等兵器；锡占二分之一的，是鉴燧之齐，适宜于制作金属镜子等器物。

4. The Bronze Caster*

Among the crafts that melt metals, the official Zhu Shi 筑氏 is responsible for bronze alloys with high tin content, and the official Ye Shi 冶氏 for bronze alloys with low tin content. The official Fu Shi 凫氏 manufactures musical instruments from bronze; the official Li Shi 栗氏 measurement devices; the official Duan Shi 段氏 farm tools; and the official Tao Shi 桃氏 weapons.

There are six different bronze alloys. Copper and tin with a ratio 6:1 are used for the manufacture of bells and ritual sacrifice vessels called *ding* 鼎;[82] in the ratio 5:1 for the manufacture of hatchets and axes; in the ratio 4:1 for the manufacture of lances and halberds; in the ratio 3:1 for the manufacture of large knives; in the ratio 5:2 for the manufacture of knives to cut bamboo strips for books and of lethal arrows; and in the ratio 2:1 for the manufacture of plane and concave mirrors.

* See commentaries: Bronze Technology and Bronze Objects.

82 *Ritual sacrifice vessels ding* 鼎 – The ritual sacrifice vessel of type *ding* 鼎 is made from bronze and generally had four legs so that it could easily be put on a fire to cook the sacrificial food.

4. Der Bronzegießer*

Bei den Handwerken zum Schmelzen von Metallen ist der Beamte Zhu Shi 筑氏 für die Bronzelegierungen mit hohem Zinnanteil und der Beamte Ye Shi 冶氏 für die Bronzelegierungen mit niedrigem Zinnanteil verantwortlich. Der Beamte Fu Shi 凫氏 stellt Musikinstrumente aus Bronze her, der Beamte Li Shi 栗氏 Messgeräte, der Beamte Duan Shi 段氏 landwirtschaftliche Geräte und der Beamte Tao Shi 桃氏 Waffen her.

Es gibt sechs verschiedene Bronzelegierungen. Kupfer und Zinn im Verhältnis 6:1 wird zur Herstellung von Glocken und rituellen Opfergefäßen *ding* 鼎[83] benutzt, im Verhältnis 5:1 zur Herstellung von Beilen und Äxten, im Verhältnis 4:1 zur Herstellung von Lanzen und Hellebarden, im Verhältnis 3:1 zur Herstellung von großen Messern, im Verhältnis 5:2 zur Herstellung von Messern zum Schneiden von Bambusstreifen für Bücher und tödlichen Pfeilen und im Verhältnis 2:1 zur Herstellung von ebenen und Hohlspiegeln benutzt.

5AC. 筑氏为削

筑氏为削，长尺博寸，合六而成规。欲新而无穷，敝尽而无恶。

冶氏为杀矢，刃长寸，围寸，铤十之，重三垸。戈广二寸，内倍之，胡三之，援四之。已倨则不入，已句则不决。长内则折前，短内则不疾，是故倨句外博。重三锊。戟广寸有半寸，内三之，胡四之，援五之。倨句中矩，与刺重三锊。

桃氏为剑，腊广二寸有半寸，两从半之。以其腊广为之茎围，长倍之，中其茎，设其后。参分其腊广，去一以为首广，而围之。身长五其茎长，重九锊，谓之上制，上士服之。身长四其茎长，重七锊，谓之中制，中士服之。身长三其茎长，重五锊，谓之下制，下士服之。

---

* Siehe Kommentare: Bronze Technology und Bronze Objects.

83 *Rituelle Opfergefäße ding* 鼎 – Das rituelle Opfergefäß des Typs ding 鼎 ist aus Bronze hergestellt und hat im Allgemeinen vier Beine, so dass man es leicht auf ein Feuer stellen kann, um die Opferspeise zu kochen.

5MC. 筑氏为削

筑氏制作削（书刀），[84] 长度为一尺，宽度为一寸。六把削合起来，正好围成一个圆。好的削应该锋利程度永远如新，而且即使用到最后，已经磨损得很厉害了，内部也没有瑕疵。

冶氏制作杀矢（箭头），箭簇长一寸，周长一寸，铤（箭头尾部用于插入箭杆中的部分）长一尺，重三垸。[85] 戈的宽度是二寸；内的长度是其二倍，四寸；胡长是其三倍，六寸；援是其四倍，八寸。援和胡之间的角度太钝，战斗中用前锋啄人就不容易啄入；角度太锐，战斗中不能割断目标。内若太长，前面的援容易折断；太短，使用起来气势不猛，因此援和胡之间的夹角应该稍微大于直角。戈的重量是三锊。戟的宽度一寸半，内的长度是其三倍，四寸半；胡的长度是其四倍，六寸；援的长度是其五倍，七寸半。戟的援和胡之间的角度应该正好等于直角，包括最前端的刺在内，戟重三锊。

桃氏制作剑，剑的两面刃（即所谓的"腊"）平面间宽二寸半，剑两面隆起的剑脊厚是其宽度的一半，即一寸二分五厘。剑柄（即所谓的 "茎"）周长与剑的 "腊广" 相等，亦为二寸半；剑柄长则加倍，为五寸，在剑柄的中部设有凸起的环箍（即所谓的"后"）。以腊广的三分之二作为剑柄末端圆盘（即所谓的 "首"）的周长。剑的长度是剑柄长度的五倍，为二尺 5 寸，重九锊，[86] 这样的剑为大号剑，供大个子佩戴；剑的长度是剑柄长度的四倍，为二尺，重七锊，这样的剑为中号剑，供中等身高人士佩戴；剑的长度是剑柄长度的三倍，为一尺五寸，重五锊，这样的剑为小号剑，供矮个子佩戴。

5. Bronze Weapons*

The official Zhu Shi 筑氏 manufactures knives for cutting bamboo strips. They are one Chi 尺 (19.7 cm) long and one Cun 寸 (2.0 cm) wide. Six such knives form a circle.[87] It is desirable that they are always sharp, like a new knife. If they are worn, the bronze is still unchanged.

84 削，书刀。先秦时没有纸笔，人们用削刻字。到了汉代，有了纸张，人们用书刀切割，故把削称作书 刀。

85 此处 "垸" 是《考工记》中的一个重量单位，具体大小不详。清代戴震认为垸即"锾"。 1锾等于 160 铢。

86 锊，古代重量单位，1锊等于160 铢。

* See commentary: Bronze Objects.

87 *Six such knives form a circle* – The bronze knives for cutting the bamboo strips that were used for writing are bent like a circular arc, with 60° centre angle. Therefore, six such knives put together can form a circle.

The official Ye Shi 冶氏 manufactures lethal arrows. The arrow tips are one Cun 寸 (2.0 cm) long, their circumference is one Cun 寸 and the shaft is one Chi 尺 (19.7 cm) long. Their weight is three Huan 锾 (17.9 g).

The lances are two Cun 寸 (3.9 cm) wide. The length of the knife *nei* 内 is twofold, that is, four Cun 寸 (7.9 cm); the length of the knife *hu* 胡 is threefold, that is, six Cun 寸 (11.8 cm); and the length of the knife *yuan* 援 is fourfold, that is, eight Cun 寸 (15.7 cm). If the angle between the knives *yuan* 援 and *hu* 胡 is too wide, the lance is not effective. If the angle between the knives *yuan* 援 and *hu* 胡 is too narrow, the lance is not deadly. If the knife *nei* 内 is too long, it can easily break off; if the knife *nei* 内 is too short, it cannot develop sufficient impact. Therefore, the knife *yuan* 援 must show a little bit upwards. The lance weighs three Lüe 锊 (248 g).[88]

A halberd is one and a half Cun 寸 (2.9 cm) wide. The length of the knife *nei* 内 is threefold, that is, four and a half Cun 寸 (8.8 cm). The length of the knife *hu* 胡 is fourfold, that is, six Cun 寸 (11.8 cm). The length of the knife *yuan* 援 is fivefold, that is, seven and a half Cun 寸 (14.7 cm). The knives *yuan* 援 and *hu* 胡 include a right angle. Together with the bayonet, the lance head weighs three Lüe 锊 (248 g).

The official Tao Shi 桃氏 manufactures swords. The width between the two cutting edges is two and a half Cun 寸 (4.9 cm). The width between the two cutting edges equals the circumference of the sword hilt, and the length of the sword hilt equals twofold its circumference. The raised rings on the hilt are in its middle. Two thirds of the width between the two cutting edges provide the diameter of the round hilt end. The length of the sword's body is fivefold the hilt length. It weighs nine Lüe 锊 (744 g), is called first class and is carried by high dignitaries. A sword that weighs seven Lüe 锊 (579 g) is called a medium sword, and is carried by medium dignitaries. A sword the length of whose body is threefold the hilt length weighs five Lüe 锊 (414 g). It is called a sword of third class, and is carried by low officials.

5. Bronzewaffen*

Der Beamte Zhu Shi 筑氏 stellt Messer zum Schneiden von Bambusstreifen her. Sie sind ein Chi 尺 (19,7 cm) lang und ein Cun 寸 (2,0 cm) breit. Sechs

88 *The lance weighs three Lüe 锊 (248 g)* – This weight seems to be too little.

* Siehe Kommentar: Bronze Objects.

solche Messer bilden zusammen einen Kreis.[89] Man möchte, dass sie immer scharf wie ein neues Messer bleiben. Sind sie abgenutzt, ist die Bronze aber unverändert.

Der Beamte Ye Shi 冶氏 stellt die tödlichen Pfeile her. Die Pfeilspitzen sind ein Cun 寸 (2,0 cm) lang, ihr Umfang beträgt ein Cun 寸, und der Schaft ist ein Chi 尺 (19,7 cm) lang. Sie sind drei Huan 锾 (17,9 g) schwer.

Die Lanzen sind zwei Cun 寸 (3,9 cm) breit. Die Länge des Messers *nei* 内 ist das Doppelte, das heißt vier Cun 寸 (7,9 cm); die Länge des Messers *hu* 胡 ist das Dreifache, das heißt sechs Cun 寸 (11,8 cm), und die Länge des Messers *yuan* 援 ist das Vierfache, das heißt acht Cun 寸 (15,7 cm). Wenn der Winkel zwischen den Messern *yuan* 援 und *hu* 胡 zu groß ist, ist die Lanze nicht wirksam. Ist der Winkel zwischen den Messern *yuan* 援 und *hu* 胡 zu klein, ist die Lanze nicht tödlich. Ist das Messer *nei* 内 zu lang, kann es leicht abbrechen; ist das Messer *nei* 内 zu kurz, kann man nicht genügend Wucht entwickeln. Darum muss das Messer *yuan* 援 ein wenig nach oben zeigen. Die Lanze wiegt drei Lüe 锊 (248 g).[90]

Eine Hellebarde ist anderthalb Cun 寸 (2,9 cm) breit. Die Länge des Messers *nei* 内 ist das Dreifache, das heißt viereinhalb Cun 寸 (8,8 cm). Die Länge des Messers *hu* 胡 ist das Vierfache, das heißt sechs Cun 寸 (11,8 cm). Die Länge des Messers *yuan* 援 ist das Fünffache, das heißt siebeneinhalb Cun 寸 (14,7 cm). Die Messer *yuan* 援 und *hu* 胡 schließen einen rechten Winkel ein. Zusammen mit dem Bajonett wiegt der Lanzenkopf drei Lüe 锊 (248 g).

Der Beamte Tao Shi 桃氏 stellt Schwerter her. Die Breite zwischen den beiden Schneiden beträgt zweieinhalb Cun 寸 (4,9 cm). Die Breite zwischen den beiden Schneiden ist gleich dem Umfang des Schwertgriffs, und die Länge des Schwertgriffs ist gleich dem Doppelten dieses Umfangs. Die erhabenen Ringe auf dem Griff befinden sich in seiner Mitte. Zwei Drittel der Breite zwischen den beiden Schneiden ergeben den Durchmesser des runden Griffendes. Die Länge des Schwertkörpers ist das Fünffache der Grifflänge. Ein Schwert, das neun Lüe (744 g) wiegt, heißt Schwert erster Klasse und wird von hohen Würdenträgern getragen. Ein Schwert, das sieben Lüe 锊 (579 g) wiegt, heißt mittleres Schwert und wird von den mittleren Würdenträgern getragen. Ein Schwert, bei dem die Länge des Schwertkörpers das Dreifache der Grifflänge

89 *Sechs solche Messer bilden zusammen einen Kreis* – Die Bronzemesser zum Schneiden der Bambusstreifen, die zum Schreiben benutzt wurden, waren wie ein Kreisbogen mit einem Zentriwinkel von 60° gebogen. Deshalb konnten sechs derartige Messer zusammen einen Kreis bilden.

90 *Die Lanze wiegt drei Lüe* 锊 *(248 g)* – Dieses Gewicht scheint zu klein zu sein.

beträgt, wiegt fünf Lüe 锊 (414 g). Es heißt Schwert dritter Klasse und wird von den niederen Beamten getragen.

6AC. 凫氏为钟

凫氏为钟，两栾谓之铣，铣间谓之于，于上谓之鼓，鼓上谓之钲，钲上谓之舞，舞上谓之甬，甬上谓之衡，钟县谓之旋，旋虫谓之干，钟带谓之篆，篆间谓之枚，枚谓之景，于上之攠谓之隧。十分其铣，去二以为钲。以其钲为之铣间，去二分以为之鼓间。以其鼓间为之舞修，去二分以为舞广。以其钲之长为之甬长，以其甬长为之围。参分其围，去一以为衡围。参分其甬长，二在上，一在下，以设其旋。薄厚之所震动，清浊之所由出，侈弇之所由兴，有说:钟已厚则石，已薄则播，侈则柞，弇则郁，长甬则震。是故大钟十分其鼓间，以其一为之厚；小钟十分其钲间，以其一为之厚。钟大而短，则其声疾而短闻；钟小而长，则其声舒而远闻。为遂,[91] 六分其厚，以其一为之深，而圜之。

6MC. 凫氏为钟

凫氏制作钟，钟下面的两角（两栾）叫做铣，两铣之间叫做于，于的上方供敲击的地方叫做鼓，鼓上部的钟体叫做钲，钲的上顶面叫舞，舞上面的悬柄叫甬，甬的上端面叫衡，用来悬吊钟的环叫旋，旋上面的纽叫干，钟上面的纹饰叫篆，篆之间凸出来的乳钉叫枚，枚又称作景，于上面供磨锉以调声调的地方叫隧。把铣的长度分成十份，去掉其中的二份，剩余的五分之四为钲的长度，把钲的长度作为两铣之间的宽度，把铣的长度的十份再去掉二份，剩余的五分之三作为两鼓之间的宽度，把两鼓之间的宽度作为舞的纵长，把铣的长度的十份继续去掉二份，剩余的铣长的五分之二作为舞的横广。以钲的长度作为甬长，以甬的长度作为其周长，以甬的周长的三分之二作为衡的周长。在甬从上到下的三分之二处设置钟环，以便悬挂钟。钟体的厚薄，决定了其受到震动时所发出的声音音调的高低；钟口的外侈和内敛的阔狭，对其也有影响。这些，有具体的说法:钟体太厚，敲起来就像击石，声音发不出来；钟体太薄，声音会显得散而不聚；钟口张得大，声音响而外播，余音嘈杂；钟口过于内敛，声音郁闷，不够悠扬；甬太长了，声音发颤而不正。因此对于大钟，钟体厚度应为其鼓间的十分之一；对于小钟，钟体厚度应为其钲间的十分之一。大而短的钟声音尖锐短暂，传播距离近；小而长的钟声音舒扬，传播距离远。制作钟口内侧供调音用的隧时，应使其面为圆弧形，高度为壁厚的六分之一。

---

91 即本段前文之"隧"字。

6. Bronze Bells*

The official Fu Shi 凫氏 casts bells. The two pointed ends of the bell are called *xian* 銑. Between the tips *xian* 銑 is the bell edge, *yu* 于. The spot where the bell is rung is called *zheng* 鉦. The top of the bell above the region *zheng* 鉦 is called *wu* 舞. The grip of the bell above the upper closing is called *yong* 甬. The end of the grip is called *heng* 衡. The ring located on the grip for hanging the bell is called *xuan* 旋. The thickening on the grip is called *gan* 幹. The decorations on the bell body in the region *zheng* 鉦 are called *zhuan* 篆. The protruding studs between the rows of decorations are called *mei* 枚; they are also called *jing* 景. The ground off region on the bell's edge is called *sui* 隧. With regard to the bell's body, four fifths of the length of the edge of the pointed end *xian* 銑 provide the distance between the two locations *gu* 鼓, where the bell is rung. The distance between the two locations *gu* 鼓 provides the vertical length of the bell's upper closing *wu* 舞, whereas two fifths of the length of the edge to the pointed end *xian* 銑 provide the width of the upper closing *wu* 舞. The length of the region *zheng* 鉦 equals the length of the grip *yong* 甬. The length of the grip equals its circumference. Two thirds of the (lower) grip's circumference provide the grip's circumference on the upper end. At a height of one third of the grip's length from below, the bell ring is located. The vibrations of the bell depend on the thickness of the bell's wall, and from this stem the high and the low tones of the bell.[92] Also, the size of the bell's opening, whether narrow or wide, exerts an influence, which is understandable. If the bell wall is too thick, it sounds as if one were hitting a stone. If the bell wall is very thin, the bell resounds. If the bell opening is very large, it sounds loudly. If the bell opening is small, it sounds suppressed. If the bell's grip is too long, the bell tone vibrates. Therefore, the wall thickness for large bells is chosen in such a way that it is one tenth of the distance between the two locations *gu* 鼓 where the bell is rung. For small bells, the wall thickness equals one tenth of the distance between the two regions *zheng* 鉦, where they border on the upper closing *wu* 舞. If the bell body is large and short, the tone is violent, but expires quickly and is not heard far away. If the bell body is small and long, the tone is calm and long lasting

---

* See commentary: Musical Instruments.

92 *From this stem the high and the low tones of the bell* – The "Kao Gong Ji" reflects the empiric knowledge that the frequency f of the tone emitted by the bell depends on the thickness t of the wall and the opening diameter of the bell. If one regards bells with an elliptic cross section, where the longer radius of the ellipse is b, then the proportionality $f \sim t/b^2$ is valid. [Chen Tong 陈通, *Zhong de zhendong he xingxi bianzhong* 钟的振动和刑 xi 编钟, *Yongle dazhong de shengxue texing* 永乐大钟的声学特性, Ziran kexue nianjian 自然科学年鉴, Shanghai, Shanghai kexue jishu chubanshe 上海科学技术出版社, 1981].

and is heard far away. The depth of the circular region *sui* 隧, ground off on the inside of the bell, amounts to one sixth of the wall thickness.

6. Bronzeglocken*

Der Beamte Fu Shi 凫氏 gießt Glocken. Die beiden spitzen Enden der Glocke heißen *xian* 銑. Zwischen den Spitzen *xian* 銑 liegt der Glockenrand *yu* 于. Der Ort, an dem die Glocke geschlagen wird, heißt *zheng* 鉦. Der obere Abschluss der Glocke über dem Bereich *zheng* 鉦 heißt *wu* 舞. Der Griff der Glocke über dem oberen Abschluss heißt *yong* 甬. Das Ende des Griffs heißt *heng* 衡. Der am Griff befindliche Ring zum Aufhängen der Glocke heißt *xuan* 旋. Die Verdickung am Griff heißt *gan* 幹. Die Verzierungen auf dem Glockenkörper im Bereich *zheng* 鉦 heißen *zhuan* 篆. Die vorstehenden Zapfen zwischen den Reihen der Verzierungen heißen *mei* 枚, sie werden auch *jing* 景 genannt. Der am Glockenrand abgeschliffene Bereich heißt *sui* 隧. Bezüglich des Glockenkörpers bilden vier Fünftel der Länge der Kante zur Spitze *xian* 銑 die Länge des Bereichs *zheng* 鉦, die zugleich dem Abstand zwischen den Spitzen *xian* 銑 entspricht. Drei Fünftel der Länge zwischen den Spitzen *xian* 銑 ergeben den Abstand zwischen den beiden Orten *gu* 鼓, an denen die Glocke geschlagen wird. Der Abstand zwischen den beiden Orten *gu* 鼓 ergibt die senkrechte Länge des oberen Abschlusses *wu* 舞 der Glocke, wobei zwei Fünftel der Länge der Kante zur Spitze *xian* 銑 die Breite des oberen Abschlusses *wu* 舞 liefert. Die Länge des Bereichs *zheng* 鉦 ist gleich der Länge des Griffes *yong* 甬. Die Länge des Griffes ist gleich seinem Umfang. Zwei Drittel des (unteren) Griffumfangs ergeben den Umfang des Griffes am oberen Ende. In einer Höhe von einem Drittel der Grifflänge von unten befindet sich der Glockenring. Die Schwingungen der Glocke hängen von der Dicke der Glockenwand ab, und von daher rührt der hohe und der tiefe Ton der Glocke.[93] Auch die Größe der Glockenöffnung, ob sie eng oder weit ist, übt einen Einfluss aus, was verständlich ist. Wenn die Glockenwand zu dick ist, klingt sie, als ob man gegen einen Stein schlägt. Wenn die Glockenwand sehr dünn ist, tönt die Glocke. Ist die

* Siehe Kommentar: Musical Instruments.

93 *von daher rührt der hohe und der tiefe Ton der Glocke* – Das „Kao Gong Ji" widerspiegelt das empirische Wissen, dass die Frequenz f des von der Glocke ausgesendeten Tons von der Dicke t der Wand und dem Öffnungsdurchmesser der Glocke abhängt. Wenn man Glocken mit elliptischem Querschnitt betrachtet, bei denen der längere Radius der Ellipse b ist, dann gilt die Proportionalität $f \sim t/b^2$. [Chen Tong 陈通, *Zhong de zhendong he jingli bianzhong* 钟的振动和刑 xi 编钟, *Yongle dazhong de shengxue texing* 永乐大钟的声学特性, Ziran kexue nianjian 自然科学年鉴 Shanghai, Shanghai kexue jishu chubanshe 上海科学技术出版社, 1981].

Glockenöffnung sehr groß, klingt sie laut. Ist die Glockenöffnung klein, klingt sie unterdrückt. Wenn der Glockengriff zu lang ist, vibriert der Glockenton. Darum wird die Wanddicke bei großen Glocken so gewählt, dass sie gleich einem Zehntel des Abstands zwischen den beiden Orten *gu* 鼓 ist, an denen die Glocke geschlagen wird. Bei kleinen Glocken ist die Wanddicke gleich einem Zehntel des Abstands zwischen den beiden Bereichen *zheng* 鉦, wo sie an den oberen Abschluss *wu* 舞 grenzen. Ist der Glockenkörper groß und kurz, ist der Ton heftig, erlischt aber schnell und wird nicht weit gehört. Ist der Glockenkörper klein und lang, ist der Ton ruhig und lang und anhaltend und wird weit gehört. Die Tiefe des auf der Innenseite der Glocke befindlichen abgeschliffenen kreisförmigen Bereichs *sui* 隧 beträgt ein Sechstel der Wanddicke.

7AC. 栗氏为量

栗氏为量，改煎金、锡则不耗，不耗然后权之，权之然后准之，准之然后量之。量之以为鬴，深尺，内方尺而圜其外，其实一鬴；其臀一寸，其实一豆；其耳三寸，其实一升。重一钧，其声中黄钟之宫。概而不税。其铭曰： “时文思索，允臻其极。嘉量既成，以观四国。永启厥后，兹器维则。” 凡铸金之状；金与锡，黑浊之气竭，黄白次之。黄白之气竭，青白次之。青白之气竭，青气次之，然后可铸也。

7MC. 栗氏为量

栗氏制作量器，反复熔冶铜、锡，当铜、锡熔液中的杂质被煎炼殆尽，反复熔冶铜锡重量也不再减少时，就可以称量所需要的铜锡溶液重量，同时校准用来筑造的型范，校准后就可以将铜锡熔液注入型范，这就得到了量器鬴。鬴是圆桶型量器，深一尺，口径正好容纳下一个边长为一尺的正方形，其容积是一鬴；底部容器开口向下，内深一寸，容积是一豆；两侧的鬴耳深三寸，容积是一升。鬴重一钧，敲击它时发出黄钟的宫音。这一量器是作为其他量器的标准来用的，不是用来收税赋的。[94]上面刻有铭文，铭

94 《周礼注疏》引《郑志》，“赵商问：‘栗氏为量，概而不税，《廛人职》有税何？’ 答曰：‘官量不税。’若然，此官量镇在市司，所以勘诸廛之量器以取平，非是寻常所用，故不税。” 按：计量是维持国家机 器正常运转的技术保障，而国家机器运转需要有税收的保障，古代税收多为实物缴纳，这就必须用到 度量衡。但标准器是用来确定一般所用的度量衡是否符合标准，而不是直接用来测量的，所以 “栗氏 量” 的重要使命是 “概而不税”，因为它是当时人精心设计出来的标准器，而不是一般的量器。

文写到："文德之君，思求为民众提供标准，而制作此器，寓诚信于中，达到极致。[95]嘉量制成，传示四方，永启子孙，遵行此器标准，维系准则。"

凡是冶炼青铜，用铜与锡作为原料，一开始冶炼时冒出黑浊的气体，然后是黄白色的气体。黄白色的气体消失后，冒出的是青白色的气体。青白色气体消失后，冒出的青色气体，这时就可以浇铸器物了。

7. Measuring Standards*

The official Li Shi 栗氏 manufactures measuring devices. First, copper and tin are melted and refined until the admixtures vanish. Then, the required amount of copper and tin is weighed, and the volume of the cast material is determined by dipping it into water and weighing the replaced water.[96] Finally, the volume vessel Fu 鬴 is cast. The main body of this volume vessel is a cylinder with a height of one Chi 尺 (19.7 cm). Inside a square is inscribed with an edge length of one Chi, and outside the vessel is round. This yields a volume of one Fu 鬴 (12.8 l). The foot of the vessel has a height of one Cun 寸 (2.0 cm) and delivers the volume of one Dou 豆 (0.8 l). The small vessels that are attached like ears to both sides of the main vessel have a height of three Cun 寸 (5.9 cm); their volume is one Sheng 升 (0.2 l). The weight of the volume vessel Fu 鬴 is one Jun 钧 (6.165 kg). If one rings the vessel, it emits the tone *Gong* 宫 of the standard pipe *Huangzhong* 黄鐘.[97] When measuring the volume of grain, the

95 《周礼注疏》郑玄注云："时，是也。允，信也。臻，至也。极，中也。言是文德之君，思求可以为民 法者，而作此量，信至于道之中。"

* See commentary: Metrology.

96 *The volume of the cast material is determined by dipping it into water and weighing the replaced water* – It is not clear whether the volume of the cast bronze or the volume of the copper and tin ingots are determined.

97 *If one rings the vessel, it emits the tone Gong* 宫 *of the standard pipe Huangzhong* 黄种 – The tone *Gong* 宫 belongs to the five-tone scale consisting of the tones *Gong* 宫, *Shang* 商, *Jue* 角, *Zhi* 徵 and *Yu* 羽. The standard pipe *Huangzhong* 黄鐘 belongs to a set of twelve standard pipes with different lengths. The sizes of each standard pipe are modified by the five tones. Therefore, there exists a standard pipe *Huangzhong Gong* 黄鐘宫 that, according to "Yue ling zhang ju" 月令章句 by Cai Yong 蔡邕 (Eastern Han dynasty) has a length of nine Cun 寸, a diameter of three Cun 寸 and a circumference of nine Cun 寸. According to the laws of acoustics, the corresponding pipe emits a frequency of f = 693.5 Hz. The description of the measurement standard Fu shows that in China, on the border between the Chunqiu 春秋 and the Zhanguo 战国 period, the idea of a tone standard already existed. However, it is unclear how the emission of a defined tone on the described vessel could be realised.

grain is levelled with a ruler, but the use of this volume vessel is not intended for imposing taxes. The inscription on the vessel runs: "The virtuous ruler worries about the people. Therefore he made this measuring device, which creates deep confidence. The standard device was successfully manufactured and proclaimed in all four directions. Forever it should be handed down to the descendants so that this device is preserved like a law."

During melting and casting of the bronze, the following states are observed: When copper and tin melt, dark mist rises. When the dark mist vanishes, yellowish-white mist rises. When the yellowish-white mist vanishes, bluish-white mist rises. When the bluish-white mist vanishes, bluish mist rises. Then one can cast the bronze.[98]

7. Messnormale*

Der Beamte Li Shi 栗氏 stellt Messgeräte her. Kupfer und Zinn werden zuerst geschmolzen und raffiniert, bis keine Beimengungen mehr vorkommen. Dann wiegt man die benötigte Menge von Kupfer und Zinn, und das Volumen des Gussmaterials wird durch Eintauchen in Wasser und Wiegen des verdrängten Wassers bestimmt.[99] Schließlich wird das Volumengefäß Fu 鬴 gegossen. Der Hauptkörper dieses Volumengefäßes ist ein Zylinder mit einer Höhe von einem Chi 尺 (19,7 cm). Innen ist ein Quadrat mit einer Kantenlänge von einem Chi 尺 eingeschrieben, und außen ist das Gefäß rund. Das ergibt ein Volumen von einem Fu 鬴 (12,8 l). Der Fuß des Gefäßes hat eine Höhe von einem Cun 寸 (2,0 cm) und bildet ein Volumen von einem Dou 豆 (0,8 l). Die wie Ohren an beiden Seiten des Hauptgefäßes angebrachten kleinen Gefäße haben eine Höhe von drei Cun 寸 (5,9 cm), ihr Volumen beträgt ein Sheng 升 (0,2 l). Das Gewicht des Volumengefäßes Fu 鬴 beträgt ein Jun 钧 (6,165 kg). Wenn man

98 *Then one can cast the bronze* – The flame-colour changes of the fluid bronze indicate the process of its refinement. In the beginning, the dark colours stem from the mixture of oxides, sulfates and metal ions. When the zinc boils above 907 °C, the emerging ZnO produces a whitish vapour that, due to the copper, additionally gets a yellowish colour. Above 1200 °C, when the green colour dominates due to the copper, the refinement is finished and the bronze can be cast. In this way, the flame colours provided criteria for the refinement process.

* Siehe Kommentar: Metrology.

99 *das Volumen des Gussmaterials wird durch Eintauchen in Wasser und Wiegen des verdrängten Wassers bestimmt* – Es ist unklar, ob das Volumen der gegossenen Bronze oder das Volumen der Kupfer- und Zinningots bestimmt wurde.

das Gefäß anschlägt, gibt es den Ton *Gong* 宫 der Standardpfeife *Huangzhong* 黃鐘 ab.[100] Beim Messen des Volumens von Korn wird das Korn mit einem Lineal glatt gestrichen, aber die Verwendung dieses Volumengefäßes ist nicht für das Eintreiben von Steuern vorgesehen. Die Inschrift auf dem Gefäß lautet: „Der tugendhafte Herrscher sorgt sich um das Volk. Deshalb schuf er dieses Messgerät, das tiefes Vertrauen schafft. Das Standardgerät wurde erfolgreich hergestellt und in allen vier Himmelsgegenden verkündet. Für immer soll es an die Nachkommen weitergegeben werden, damit dieses Gerät wie ein Gesetz bewahrt wird."

Beim Schmelzen und Gießen der Bronze werden folgende Zustände beobachtet: Wenn Kupfer und Zinn schmelzen, steigen dunkle Nebel auf. Wenn die dunklen Nebel verschwunden sind, steigen gelblich-weiße Nebel auf. Wenn die gelblich-weißen Nebel verschwunden sind, steigen bläulich-weiße Nebel auf. Wenn die bläulich-weißen Nebel verschwunden sind, steigen bläuliche Nebel auf. Dann kann man die Bronze gießen.[101]

---

100 *Wenn man das Gefäß anschlägt, gibt es den Ton Gong* 宫 *der Standardpfeife Huangzhong* 黄种 *ab* – Der Ton *Gong* 宫 gehört zu einer Fünf-Tonleiter, die aus den Tönen *Gong* 宫, *Shang* 商, *Jue* 角, *Zhi* 徵 und *Yu* 羽 besteht. Die Standardpfeife *Huangzhong* 黄种 gehört zu einem Satz von zwölf Standardpfeifen unterschiedlicher Längen. Die Abmessungen jeder Standardpfeife werden durch die fünf Töne modifiziert. Deshalb gibt es eine Standardpfeife *Huangzhong Gong* 黄种宫, die entsprechend dem Werk „*Yue ling zhang ju*" 月令章句 von Cai Yong 蔡邕 (Östliche Han 东汉-Dynastie) eine Länge von neun Cun, einen Durchmesser von drei Cun und einen Umfang von neun Cun hat. Nach den Gesetzen der Akustik erzeugt die entsprechende Pfeife eine Frequenz von f = 693,5 Hz. Die Beschreibung des Messnormals Fu 鬴 zeigt, dass in China an der Grenze zwischen der Chunqiu 春秋- und der Zhanguo 战国-Periode bereits die Idee eines Ton-Standards existierte. Aber es bleibt unklar, wie die Abgabe eines definierten Tons an dem beschriebenen Gefäß realisiert werden konnte.

101 *Dann kann man die Bronze gießen* – Die Änderungen der Flammenfarben der flüssigen Bronze zeigen den Prozess ihrer Raffination an. Am Anfang rührt die dunkle Farbe von einem Gemisch von Oxiden, Sulfaten und Metallionen. Wenn Zink oberhalb 907 °C siedet, liefert das entstehende ZnO einen weißlichen Dampf, der durch das Kupfer zusätzlich eine gelbliche Farbe erhält. Wenn die grüne Farbe oberhalb 1200 °C aufgrund des Kupfers dominiert, ist die Raffination abgeschlossen, und die Bronze kann gegossen werden. Auf diese Weise lieferten die Flammenfarben ein Kriterium über den Raffinationsprozess.

8AC. 段氏阙。

8MC. 段氏（已阙）

8. Agricultural Tools (Missing)

8. Landwirtschaftliche Geräte (fehlt)

9AC. 函人为甲

函人为甲，犀甲七属，兕甲六属，合甲五属。犀甲寿百年，兕甲寿二百年，合甲寿三百年。凡为甲，必先为容，然后制革。权其上旅与其下旅，而重若一，以其长为之围。凡甲，锻不挚则不坚，已敝则桡。凡察革之道，眡其钻空，欲其窓也；眡其里，欲其易也；眡其朕，欲其直也；櫜之，欲其约也；举而眡之，欲其丰也；衣之，欲其无齘也。眡其钻空而窓，则革坚也；眡其里而易，则材更也；眡其朕而直，则制善也。櫜之而约，则周也；举之而丰，则明也；衣之无齘，则变也。

9MC. 函人为甲

函人制作皮甲，用犀牛皮制作的甲以 7 组革连缀而成；用雌犀牛皮制作的甲，以六组皮革连缀而成；削去皮革表皮内侧的肉质，取两张表皮合在一起制成的合甲，用五组皮革连缀而成。犀牛皮甲寿命一百年，雌犀牛皮甲寿命二百年，而合甲的寿命则达到三百年。凡是制作皮甲，必须先按照人体的大小长短做好样式，然后才能裁制皮革。要分别称量腰上部分和腰下部分的甲，使其重量相等，以甲的长度作为其腰围。凡是制作皮甲，用锤敲打得不细致周到，革就不坚实；捶打得过分，革会受伤而翘曲。凡是检查革的质量，有一定的要领：观察其缝缀的针孔，针孔越细越好；查看它的内面，要求整洁细密；查看其缝合的甲缝，要求顺直；将其收放在甲囊里，要求它易于收放，叠放密实体积小；将其举起展开，要求它显得形体宽大；穿到身上，合身而舒适。针孔小，说明革一定很坚实；其内面整洁细密，革的质量一定很好；接缝顺直，制造的工艺一定很高超；放入甲囊中密实而体积小，说明各方面都顺妥；举起来显得宽大，革一定光润明亮；穿上合体舒适，则行动举止一定很便利。

9. Leather

The armour maker manufactures harnesses. For a harness made from rhinoceros leather,[102] seven leather pieces are tied together; for a harness from oxen leather, six leather pieces; and for a harness made of two leather layers, five leather pieces. A harness from rhinoceros leather lasts two hundred years, and a harness from two leather layers three hundred years. For the manufacture of a harness, one must first measure the body of the person and make a model of the body. After this, the leather pieces are cut. One must take care that the harness is equally heavy above and below the hip. The length of the harness equals its hip size. If the leather pieces are not hammered accurately, they are not strong. If they are hammered too heavily, the leather fibres will be destroyed and the leather pieces will buckle. The essence of processing the leather pieces is the following: One has to take care that the holes for connecting the leather pieces are as small as possible. The lining must be even and fine. The seam of the leather pieces should be straight. If the rolled-up harness is put into the harness bag, the volume should be small. If one holds up the harness with one's hands, it should look magnificent. If one has put on the harness, it should fit the body. If one sees that the holes, with which the leather pieces are tied together, are small, then the leather harness is solid. Through the even and fine leather lining, one sees the good quality. Through straight seams, one recognises good craftsmanship. If the harness fits well into the harness bag, it is manufactured carefully. If the harness looks magnificent when held up, it will win fame. If the harness fits well when it is put on, it is convenient.

9. Leder

Der Rüstungsmacher stellt Harnische her. Bei einem Nashornlederpanzer werden sieben Lederstücke, bei einem Rindslederpanzer werden sechs Lederstücke und bei einem Panzer aus zwei Lederlagen werden fünf Lederstücke zusammengebunden. Ein Nashornlederpanzer[103] hält hundert Jahre, ein Rindslederpanzer zweihundert Jahre und ein Panzer aus zwei

---

102 *Harness from rhinoceros leather* – The mention of the rhinoceros shows that at the time of the "Kao Gong Ji", the rhinoceros still existed in China. Nowadays, it lives only in India, Java, Sumatra and Africa. Moreover, in Xingping 兴平 (Shaanxi 陕西 province) in 1963, a bronze wine vessel from the Zhanguo 战国 period in the shape of a rhinoceros was excavated.

103 *Nashornlederpanzer* – Die Erwähnung des Rhinoceros zeigt, dass das Rhinoceros zur Zeit des „Kao Gong Ji" noch in China lebte. Heutzutage lebt es nur in Indien, Java, Sumatra und

Lederlagen dreihundert Jahre. Bei einer Herstellung eines Panzers muss man erst den Körper der Person abmessen und ein Körpermodell herstellen. Danach werden die Lederstücke geschnitten. Man muss dafür sorgen, dass die Panzer ober- und unterhalb der Hüfte gleich schwer sind. Die Länge des Panzers ist gleich seinem Hüftumfang. Wenn die Lederstücke nicht fein gehämmert werden, sind sie nicht fest. Werden sie zu stark gehämmert, werden die Lederfasern zerstört, und die Lederstücke wölben sich auf. Das Wesentliche bei der Verarbeitung der Lederstücke ist folgendes: Man muss darauf achten, dass die Löcher zum Verbinden der Lederstücke möglichst klein sind. Das Futter muss glatt und fein sein. Die Naht der Lederstücke soll gerade sein. Wenn der Panzer zusammengerollt in den Panzersack gesteckt wird, soll das Volumen klein sein. Wenn man den Panzer mit den Händen hochhält, soll er stattlich wirken. Wenn man den Panzer angezogen hat, soll er richtig sitzen. Wenn man sieht, dass die Löcher, mit denen die Lederstücke miteinander verbunden sind, klein sind, dann ist der Lederpanzer fest. Am glatten und feinen Lederfutter zeigt sich eine gute Qualität. An geraden Nähten erkennt man die gute Verarbeitung. Wenn der Panzer gut in den Panzersack passt, ist er sorgfältig hergestellt. Wenn der Panzer hochgehalten stattlich wirkt, bringt er Ruhm. Wenn der Panzer angezogen gut passt, ist er bequem.

10AC. 鲍人

鲍人[104]之事，望而眡之，欲其荼白也；进而握之，欲其柔而滑也；卷而抟之，欲其无迆也；眡其着，欲其浅也；察其线，欲其藏也。革欲其荼白，而疾澣之则坚；欲其柔滑而腥，脂之则需；引而信之，欲其直也。信之而直，则取材正也。信之而枉，则是一方缓一方急也。若苟一方缓一方急，则及其用之也，必自其急者先裂。若苟自急者先裂，则是以博为帴也。卷而抟之而不迆，则厚薄序也。眡其着而浅，则革信也。察其线而藏，则虽敝不甐。

Afrika. Außerdem wurde 1963 in Xingping 兴平 (Provinz Shaanxi 陕西) ein bronzenes Weingefäß in Form eines Rhinoceros aus der Zhanguo 战国-Periode ausgegraben.

104 鲍人，鞣治皮革的工匠。皮革从动物身上剥下来后，需要经过鞣治。未经鞣治的皮革称为“生皮”，“生皮”不但干后特别硬，而且其所含的蛋白质和脂肪有利于各种微生物繁殖，容易导致腐烂。皮经过鞣治，就成了“熟皮”，“熟皮”皮质柔软，抗潮防霉，坚固耐用，可以用来制作各种皮制品。

10MC. 鲍人

鲍人鞣治皮革。鞣治后的皮革，远看上去要颜色像茅草的白花一样；握在手中，要柔软光滑；将其卷紧，要两端平齐不斜；查看缝合之处，一定要浅平；查看缝线，要隐藏不露。革的颜色像茅草的白花一样，再快速地洗鞣，使鞣料渗入皮革，皮革就会很坚固；要使革柔滑润泽，涂上油脂它就会变得柔软。将其拉伸开来，希望它平直。拉伸开来很平直，就说明材料的裁取平正。伸展以后是歪斜的，则是由于一边松一边紧的缘故。如果一边松一边紧，那么到使用的时候，绷得紧的一边一定会先裂开。如果绷得紧的一侧先裂开，就需要把裂开的部分剪去，这样宽幅皮革就只能作为窄幅使用。卷紧后两端平齐不斜，说明皮革厚薄均匀。查看缝合处接缝浅平，那就说明革伸展得均匀舒展。查看缝线，它隐藏不露，这时即使革用得破旧了，缝的线也不会受到损伤。

10. The Tanner

The work of the tanner. If one looks at softly tanned leather from afar, it should look whitish, like the blossoms of cogongrass. If one touches it, it should feel soft and smooth. If one rolls it up tightly, both ends should stay even. The joint between two leather pieces should be flat. The thread that connects two leather pieces should be hidden. If the tanned leather is whitish like the blossom of cogongrass and is washed with fast movements, it is strong. The leather should be quite soft and smooth. If one coats it in fat, it becomes very soft. When pulled, the leather should stay smooth and even. Then one can sew it smoothly. When the pulled leather is not even, then one side is loose and the other side is tensed up. If one processes such a piece of leather, the tensed-up side can tear. Because such places tear first, one cannot avoid cutting them out, so that from a wide piece, one gets a narrow one. If the leather, when rolling it up, stays even, its thickness is even as well. If one sees that the seam of two leather pieces is flat, the leather will not twist. If the thread of the seam is hidden, it will hold, even if the leather is worn.

10. Der Gerber

Die Arbeit des Gerbers. Betrachtet man das weich gegerbte Leder von Weitem, muss es weißlich wie die Blüten des Süßgrases aussehen. Wenn man es anfasst, muss es sich weich und glatt anfühlen. Rollt man es fest zusammen, müssen die beiden Seiten gleichmäßig bleiben. Die Verbindungsstelle von zwei Lederstücken muss flach sein. Der Faden, der zwei Lederstücke verbindet, muss verborgen sein. Wenn das gegerbte Leder weißlich wie die Blüte des Süßgrases ist und es mit schnellen Bewegungen gewaschen wird, ist es fest. Das Leder muss ganz weich und glatt sein. Wenn man es dick mit Fett bestreicht, wird es sehr weich. Wenn man das Leder zieht, muss es glatt und eben bleiben. Dann kann man es glatt vernähen. Wenn das Leder beim Ziehen uneben ist, dann ist eine Seite lose und die andere Seite verspannt. Wenn man ein solches Lederstück verarbeitet, kann die verspannte Seite reißen. Weil solche Stellen zuerst reißen, kann man nicht umhin, sie herauszuschneiden, so dass aus einem breiten ein schmales Stück wird. Wenn das Leder beim Zusammenrollen gleichmäßig bleibt, ist auch seine Dicke gleichmäßig. Wenn man sieht, dass die Naht von zwei Lederstücken flach ist, wird sich das Leder nicht verziehen. Wenn der Faden der Naht verborgen ist, wird er halten, selbst wenn das Leder abgenutzt ist.

11AC. 韗人

韗人为皋陶，长六尺有六寸，左右端广六寸，中尺，厚三寸，穹者三之一，上三正。鼓长八尺，鼓四尺，中围加三之一，谓之鼖鼓。为皋鼓，长寻有四尺，鼓四尺，倨句磬折。凡冒鼓，必以启蛰之日。良鼓瑕如积环。鼓大而短，则其声疾而短闻；鼓小而长，则其声舒而远闻。

11MC. 韗人

韗人制作皮鼓，[105] 鼓长六尺六寸，左右两端宽六寸，中间部分宽一尺，鼓木厚三寸。鼓的中央腰部向外凸起的地方高相当于其直径的三分之一，把整个鼓木分成三段，中间腰部一段，两端各一段，鼓木均匀平直。这是一种鼓。另一种鼓叫鼖鼓，鼓长八尺，鼓面直径四尺，中间腰部外凸，高为端面直径的三分之一。还有一种鼓叫皋鼓，鼓长一丈二尺，鼓面直径四尺，中间腰部两侧折曲的夹角等于一个磬折。凡是给鼓张鼓面，一定要在二十四节气的惊蛰那天。质量精良的鼓，鼓面油漆后形成的纹理呈现出一个一个的同心圆。口径大而鼓身短，这样的鼓声调高而急促，传得不远；口径小而鼓身长，这样的鼓声音舒展，传的距离远。

105 《周礼注疏》引郑众云："皋陶，鼓木也。" 即 "皋陶" 指鼓架，这里泛指整体的鼓。

11. The Drum Maker*

The drum maker manufactures drums. Each drum slat is six Chi 尺, six Cun 寸 (129.7 cm) long. The width on both ends, right and left, is six Cun 寸 (11.8 cm), and in the middle, one Chi 尺 (19.7 cm). The thickness of the slats is three Cun 寸 (5.9 cm). The height of the buckle in the middle amounts to one third of the diameter of the drum's surface. The drum slats are arranged in three sections, where each slat must be even and straight. The drum is eight Chi 尺 (157.2 cm) high; the diameter of the drum surface is four Chi 尺 (78.6 cm). The diameter of the drum's body is one third larger than the diameter of the drum's surface. This is called the Big Drum. If one wants to manufacture a drum called *gao* 皋, its length is one Xun 寻, four Chi 尺 (235.8 cm) and the diameter of the drum's surface is four Chi 尺 (78.6 cm). The obtuse angle of the drum's body, with its bent ends is one *qingzhe* 磬折 (151°52'30"). Drums covered with leather should be manufactured on the day of the awakening of insects.[106] The lacquer patterns on the drum's leather look like many rings. If the drum is large but short, it sounds high and piercing, and its tone does not carry far. If the drum is small but long, it sounds deep and quiet, and its tone is heard far away.

11. Der Trommelmacher**

Der Trommelmacher stellt Trommeln her. Jedes Trommelholz ist sechs Chi 尺 sechs Cun 寸 (129,7 cm) lang. Die Breite an den beiden Enden rechts und links beträgt sechs Cun 寸 (11,8 cm) und in der Mitte ein Chi 尺 (19,7 cm). Die Dicke der Bretter beträgt drei Cun 寸 (5,9 cm). Die Höhe der Aufwölbung in der Mitte beträgt ein Drittel des Durchmessers der Trommelfläche. Man ordnet die Trommelhölzer in drei Abschnitten an, wobei jedes Holz eben und gerade sein muss. Die Trommel ist acht Chi 尺 (157,2 cm) hoch, der Durchmesser der Trommelfläche beträgt vier Chi 尺 (78,6 cm). Der Durchmesser des Trommelbauchs ist um ein Drittel größer als der Durchmesser der Trommelfläche. Das nennt man eine Großtrommel. Will man eine Trommel *gao* 皋 herstellen, so beträgt ihre Länge ein Xun 寻 vier Chi 尺 (235,8 cm) und der Durchmesser der Trommelfläche vier Chi 尺 (78,6 cm). Der stumpfe

* See commentary: Musical Instruments.

106 *Drums covered with leather should be manufactured on the day of the awakening of insects* – The period of the awakening of insects is one of 24 periods into which a year was divided. According to the sun calendar, the period of the awakening of insects begins around the 6th of March.

** Siehe Kommentar: Musical Instruments.

Winkel, den der Trommelbauch mit den beiden gebogenen Enden einschließt, beträgt ein Qingzhe 磬折 (151° 52'30"). Die mit Leder bespannten Trommeln muss man am Tag des Erwachens der Insekten herstellen.[107] Die Lackmuster auf dem Trommelleder sehen wie zahlreiche Ringe aus. Wenn die Trommel groß, aber kurz ist, klingt sie hoch und durchdringend, und ihr Ton wird nicht weit getragen. Wenn die Trommel klein, aber lang ist, klingt sie tief und ruhig, und ihr Ton ist weit zu hören.

12AC. 韦氏 (阙)

12MC. 韦氏（已阙）

12. Tailor of Mourning Garments (Missing)

12. Der Schneider von Trauergewändern (fehlt)

13AC. 裘氏（阙）

13MC. 裘氏（已阙）

13. Furrier (Missing)

13. Der Kürschner (fehlt)

14AC. 画缋

画缋之事，杂五色。东方谓之青，南方谓之赤，西方谓之白，北方谓之黑，天谓之玄，地谓之黄。青与白相次也，赤与黑相次也，玄与黄相次也。青与赤谓之文，赤与白谓之章，白与黑谓之黼，黑与青谓之黻，五采备谓之绣。土以黄，其象方，天时变。火以圜，山以章，水以龙，鸟兽蛇。杂四时五色之位以章之，谓之巧。凡画缋之事，后素功。

107 *Tag des Erwachens der Insekten* – Die Periode des Erwachens der Insekten ist eine der 24 Perioden, in die ein Jahr eingeteilt wurde. Nach dem Sonnenkalender beginnt die Periode des Erwachens der Insekten um den 6. März.

14MC. 画缋

画缋的事情，首先需要调配五色。东方之色是青色，南方之色是红色，西方之色是白色，北方之色是黑色，与天对应的颜色是玄色，[108] 与地对应的颜色是黄色。青色与白色相次，[109] 红色与黑色相次，玄色与黄色相次。青色与红色相间形成的条纹，叫做文；红色与白色相间形成的条纹，叫做章；白色与黑色相间形成的条纹，叫做黼；黑色与青色相间形成的条纹，叫做黻；五彩具备，叫做绣。画土用黄颜色，用方形象征大地。画天的时候随季节变化而选用不同的颜色。用圆形象征火，用獐子代表山，用龙来表现水，还要画鸟、兽、蛇作为装饰。调配好一年四季对应的不同颜色，使五色各得其位，色彩鲜明，这才叫做技巧高超。凡是画缋之类的事情，在绘好五彩之后，还要再施以白粉，以使画面光彩鲜亮。

14. Paints

The work of painters and embroiderers. The basic colours are assigned to the five directions.[110] To the east belongs blue, to the south – red, to the west – white, to the north – black. The sky is symbolised by a black-blue colour, and earth by a yellow colour. With regards to the colours of cloth, blue and white belong together, followed by red and black and by black-blue and yellow. Patterns in blue and red are called *wen* 文, patterns in red and white are called *zhang* 章, patterns in white and black are called *fu* 黼. Cross-shaped patterns in black and blue are called *fu* 黻. A complete arrangement of the colours is called *xiu* 綉 (embroidery). The earth is painted yellow, which corresponds to a cardinal direction. Depending on the season, the sky is painted with different colours. When painting the star Great Fire, it is symbolised by a round shape. When painting a mountain, it is symbolised by the river deer. When painting water, it is symbolised by the dragon. Additionally, one paints birds, wild animals and snakes. If one arranges the four seasons and the five colours in a suitable way, one gets fresh colours. This is called artistic skill. When painting,

108 "玄色" 本来即为黑色，此处 "北方谓之黑，天谓之玄"，用意何在？《周礼注疏》贾公彦对之有所解释："天玄与北方黑，二者大同小异。何者？玄黑虽是其一，言天止得谓之玄天，不得言黑天。"

109 "相次"，指相邻而渐次，即由一种颜色可以自然过渡到另一种颜色。

110 *The basic colours are assigned to the five directions* – This description reflects the idea of the five elements or changing states (*wu xing* 五行), which forms a basis of the original natural sciences. Many phenomena of nature were assigned to the five elements, and society was divided into five constituents as well. In this way, the five colours (white, green, black, red, yellow) are assigned to the five directions.

one first applies the colours, and only after this the white colour, which adds to the effect.

14. Farben

Die Arbeiten der Maler und Stickerinnen. Die Grundfarben sind den fünf Richtungen zugeordnet.[111] Zum Osten gehört blau, zum Süden rot, zum Westen weiß, zum Norden schwarz. Der Himmel wird durch die schwarzblaue Farbe und die Erde durch die gelbe Farbe symbolisiert. Bezüglich der Zuordnung der Stofffarben gehören blau und weiß zusammen, gefolgt von rot und schwarz und von schwarzblau und gelb. Muster in blau und rot heißen *wen* 文, beilförmige Muster in rot und weiß heißen *zhang* 章, Muster in weiß und schwarz heißen *fu* 黼. Kreuzförmige Muster in schwarz und blau heißen *fu* 黻. Die vollständige Zusammenstellung der Farben heißt *xiu* 綉 (Stickerei). Erde wird gelb gemalt, die zugleich eine Himmelsrichtung darstellt. Der Himmel wird je nach Jahreszeit mit einer anderen Farbe gemalt. Malt man den Stern Großes Feuer, so wird er durch die runde Form symbolisiert. Malt man einen Berg, so wird er durch das Wasserreh symbolisiert. Malt man Wasser, so wird es durch den Drachen symbolisiert, dazu malt man Vögel, wilde Tiere und Schlangen. Wenn man die vier Jahreszeiten und die fünf Farben geeignet einander zuordnet, erhält man frische Farben. Das nennt man Kunstfertigkeit. Beim Malen trägt man zuerst die Farben auf und danach erst das Weiß, was die Wirkung unterstreicht.

15AC. 钟氏

钟氏染羽，以朱湛[112]丹秫，三月而炽之，淳而渍之。三入为纁，五入为緅，七入为缁。

111 *Die Grundfarben sind den fünf Richtungen zugeordnet* – Diese Beschreibung widerspiegelt die Theorie der Fünf Elemente oder Wandlungszustände (*wu xing* 五行), die eine Grundlage der ursprünglichen Naturwissenschaften darstellte. Den Fünf Elementen wurden viele Phänomene der Natur und auch der Gesellschaft zugeordnet, die jeweils in fünf Bestandteile aufgeteilt sind. Auf diese Weise wurden die fünf Farben den fünf Richtungen zugeordnet.

112 湛，浸泡。

15MC. 钟氏

钟氏给羽毛染色，在水中浸泡朱砂和红粟米，三个月后加热，等溶液浓缩到一定程度后，就可以把羽毛浸入染色了。浸染三次染成纁色即浅红色，浸染五次成緅色即深青透红颜色，七次成缁色即黑色。

15. Dyer of Feathers

The official Zhong Shi 钟氏 dyes feathers. The feathers are placed in a solution of cinnabar and red sticky gaoliang 高粱[113] and boiled after three months. At this time, the sticky gaoliang 高粱 becomes syrupy. If the feathers are placed in this solution three times, they become light red; five times – they become dark purple and seven times – black.[114]

15. Färber von Federn

Der Beamte Zhong Shi 钟氏 färbt Federn. Die Federn werden in eine Lösung aus Zinnober und rotem klebrigen Gaoliang 高粱[115] gelegt und nach drei Monaten gekocht. Dabei wird der klebrige Gaoliang 高粱 dickflüssig. Wenn man die Federn dreimal in diese Lösung legt, werden sie hellrot, bei fünfmaligem Einlegen – dunkelviolett und bei siebenmaligem Einlegen schwarz.[116]

113 *Gaoliang* 高粱 – sorghum millet that is widely grown in northern China.

114 *If the feathers are placed in this solution three times, they become light red; five times – they become dark purple and seven times – black* – As dyeing agents the plants madder or Chinese gromwell are used; as mordant, alum is used. The colour becomes darker when the dyeing procedure is repeated several times.

115 *Gaoliang* 高粱 – Eine Sorghum-Hirse, die in Nordchina viel angebaut wird.

116 *Wenn man die Federn dreimal in diese Lösung legt, werden sie hellrot, bei fünfmaligem Einlegen – dunkelviolett und bei siebenmaligem Einlegen schwarz* – Als Färbemittel wurden die Pflanzen Krapp oder Steinsame und als Beizmittel Alaun benutzt. Die Farbe wird dunkler, wenn der Färbeprozess mehrere Male wiederholt wird.

16AC. 筐人（阙）

16MC. 筐人（已阙）

16. Dyer of Clothes and Silks (Missing)

16. Färber für Stoffe und Seiden (fehlt)

17AC. 慌氏

慌氏湅丝，以涚水沤其丝七日，去地尺暴之。昼暴诸日，夜宿诸井。七日七夜，是谓水湅。湅帛，以栏为灰，渥淳其帛，实诸泽器，淫之以蜃，清其灰而盝之、而挥之、而沃之、而盝之、而涂之、而宿之。明日，沃而盝之，昼暴诸日，夜宿诸井。七日七夜，是谓水湅。

17MC. 慌氏

慌氏练丝，用草木灰水[117] 浸泡蚕丝七天，将浸泡后的蚕丝放在离地一尺的架子上暴晒。白天在日光下暴晒，晚上将其浸泡在井水中，这样连续七天七夜，叫做水练法。练帛则需要将帛用楝木[118] 灰水充分浸泡，将其放在光滑的容器中，再加上蚌壳粉浸泡。将浸泡水沉淀、澄清，抖去沉淀在帛上的灰，重新加水，过滤，然后涂以蚌壳粉，静置过夜。第二天继续在帛上浇水，再滤去水，白天在太阳下暴晒，晚上浸泡在井水中，这样持续七天七夜，叫做帛的水练法。

17. Dyer of Silk

The official Huang Shi 慌氏 boils raw silk. He dips the silk into warm water, which is mixed with plant ashes. After seven days, the silk is dried one Chi 尺 (19.7 cm) above the earth in the sun. Every night, it is watered in well water. After seven days and nights, one obtains raw silk with the bast removed. To boil the raw silk, the leaves of the chinaberry tree are burnt to ashes.[119] A thick

117 涚水，按《周礼注疏》郑玄注，为草木灰水。

118 原文中为"栏"，据《周礼注疏》，"栏，音练。"闻人军指出："因楝叶灰水是钾溶液，呈碱性，渗透性较好，所以楝叶是传统的练丝原料。"其说甚是。

119 *The leaves of the chinaberry tree are burnt to ashes* – The ash contains potash lye, which dissolves the bast.

solution is made from these ashes, into which the silk is dipped. Limestone is dissolved in water in a smooth vessel. When the lime deposits, the water is poured out. The deposit is stirred, water is added, and again drained. After this, the limestone is spread onto the silk, and one lets it rest overnight. On the next day, the silk is rinsed in water, after which the water is drained. During the day, the silk is dried in the sun, and at night it is dipped into well water. If the silk is processed in this way for seven days and seven nights, one obtains silk from which the bast is removed by water.[120]

17. Färber für Seide

Der Beamte Huang Shi 慌氏 kocht die Rohseide. Er taucht die Seide in warmes Wasser, die mit Pflanzenasche versetzt ist. Nach sieben Tagen wird die Seide ein Chi 尺 (19,7 cm) über der Erde in der Sonne getrocknet. Nachts wird sie jeweils in Brunnenwasser gewässert. Nachdem so sieben Tage und sieben Nächte vergangen sind, erhält man die entbastete Rohseide. Zum Auskochen der Rohseide werden die Blätter des Paternosterbaums zu Asche verbrannt.[121] Aus der Asche wird eine dicke Lösung hergestellt, in die die Seide eingetaucht wird. In einem glatten Gefäß löst man Muschelkalk in Wasser auf. Wenn sich der Kalk abgesetzt hat, gießt man das Wasser ab. Die Ablagerung wird gerührt, man gibt Wasser zu und gießt das Wasser wieder ab. Danach streicht man den Muschelkalk auf die Seide und lässt sie die Nacht über ruhig liegen. Am nächsten Tag wird die Seide in Wasser gespült, worauf das Wasser abgezogen wird. Tagsüber wird die Seide in der Sonne getrocknet, und nachts wird die Seide in Brunnenwasser getaucht. Wenn man die Seide so sieben Tage und sieben Nächte behandelt, erhält man mit Wasser entbastete Seide.[122]

---

120 *One obtains silk from which the bast is removed by water* – The alternating of rinsing the silk in water and drying it in the sun is a procedure to remove the bast from the silk that is still used today.

121 *Zum Auskochen der Rohseide werden die Blätter des Paternosterbaums zu Asche verbrannt* – Die Asche enthält Kaliumlauge, die den Bast auflöst.

122 *Erhält man mit Wasser entbastete Seide* – Das abwechselnde Spülen der Seide in Wasser und Trocknen in der Sonne ist ein auch heute noch angewendetes Verfahren, um den Bast von der Seide zu entfernen.

卷下 (Part Two)

18AC. 玉人

玉人之事，镇圭尺有二寸，天子守之；命圭九寸，谓之桓圭，公守之；命圭七寸，谓之信圭，侯守之；命圭七寸，谓之躬圭，伯守之。天子执冒四寸，以朝诸侯。天子用全，上公用龙，侯用瓒，伯用将，继子男执皮帛。天子圭中必，四圭尺有二寸，以祀天；大圭长三尺，杼上终葵首，天子服之。土圭尺有五寸，以致日，以土地；祼圭尺有二寸，有瓒，以祀庙；琬圭九寸而缫，以象德；琰圭九寸，判规，以除慝，以易行。璧羡度尺，好三寸，以为度。圭璧五寸，以祀日月星辰。璧琮九寸，诸侯以享天子。谷圭七寸，天子以聘女。大璋中璋九寸，边璋七寸，射四寸，厚寸，黄金勺，青金外，朱中，鼻寸，衡四寸，有缫，天子以巡守，宗祝以前马。大璋亦如之，诸侯以聘女。瑑圭璋八寸，璧琮八寸，以覜聘。牙璋中璋七寸，射二寸，厚寸，以起军旅，以治兵守。驵琮五寸，宗后以为权。大琮十有二寸，射四寸，厚寸，是谓内镇，宗后守之。驵琮七寸，鼻寸有半寸，天子以为权。两圭五寸有邸，以祀地，以旅四望。瑑琮八寸，诸侯以享夫人。案十有二寸，枣栗，十有二列，诸侯纯九，大夫纯五，夫人以劳诸侯。璋邸射素功，以祀山川，以致稍饩。

18MC. 玉人

玉匠制玉，有各种规格与功能。镇圭长一尺二寸，天子执掌；天子颁发的圭是命圭，长九寸的命圭叫桓圭，由公爵执守；[1]长七寸的命圭叫信圭，侯爵执守；另一种长七寸的命圭叫躬圭，伯爵执守。天子用的冒长四寸，在朝会诸侯时使用。[2] 天子用纯玉，上公用玉石（玉和石之比为四比一），侯用玉和石之比为三比二的玉石，伯用玉和石各占一半的玉石。[3]接下去子爵和男爵用皮帛。天子祭天之圭共有四个，每个长一尺二寸，在其中部用丝带相系。[4]大圭长三尺，上端削成方椎形，天子服带。土圭长度一尺五寸，用来确定夏至、测量土地。用于祭祀宗庙的祼圭长一尺二寸，一头带盘。[5]琬圭长九寸，圆头，下面衬以木垫，以之象征德行。[6]琰圭长九

1 《周礼注疏》云："命圭者，王所命之圭也。朝覲执焉，居则守之。"
2 冒，通瑁。天子用瑁来检验诸侯持的圭的上端与瑁的下端是否相合。
3 原文中的"全、龙、瓒、将"，指的是不同质地的玉或玉石。译文据《周礼注疏》。
4 "必"通"縪"，指丝带。
5 《周礼注疏》云："瓒如盘，其柄用圭，有流前注。"
6 琬意为圆，指圭两端是圆形。缫指缫藉，指用跟玉大小相称的木板做的垫板。

 | DOI:10.1163/9789004416949_004

寸，一端削成圆弧尖刃，[7]象征去除邪恶，使持圭者改易劣行。玉璧长一尺宽九寸，中央孔径三寸，用来作为尺度的基准。[8]圭璧五寸，用以祭祀日月星辰。璧琮九寸，诸侯以其供献天子。圭面饰以粟纹的为谷圭，长七寸，天子用来聘女。大璋中璋长九寸，边璋长七寸，上面的射长四寸，厚一寸。〔瓒〕以黄金做勺，外镶青金，内髹朱漆，瓒一头有勺，勺的流长一寸，口径四寸，配有衬垫。天子执璋以巡守山川，大山川用大璋，中等山川用中璋，小山川用小璋。负责祭祀的官员在杀马祭山川之前要执瓒酌酒浇地，行灌礼。诸侯聘女用的大璋跟天子用的边璋规格一样。瑑圭瑑璋[9]长八寸，璧外径八寸、琮长八寸，上公之臣以之覜聘。牙璋、中璋长七寸，其射长二寸、厚一寸，供天子调动军队、治理兵事之用。琮五寸成组，王后用其为权衡之砝码。大琮长十二寸，其射长四寸，厚一寸，是内镇之宝，由王后掌管。琮七寸成组，其鼻纽高一寸半，天子用其为权衡之砝码。两个各长五寸的圭，各自有底，将其底部相对摆放，用以祭祀大地，旅祭四方。八寸大的瑑琮，诸侯以其供献国君夫人。用玉装饰的几案高一尺二寸，案上摆放枣栗，这样可摆放十二列几案。招待诸侯都摆放九列，招待大夫都摆放五列，这是国君夫人招待诸侯的规格。玉璋有基座，璋首削尖，璋身不加雕饰，这样的玉璋用于祭祀山川，也用于给宾客致送食物。

Part 2: 18. The Jade Polisher*

The work of the jade polisher. The jade sceptre *zhen gui* 鎮圭, which is reserved for the son of heaven, is one Chi 尺, two Cun 寸 (23.6 cm) long. A nine Cun 寸 (17.7 cm) long jade sceptre, which is kept by the princes, is called *huan gui* 桓圭. A seven Cun 寸 (13.8 cm) long jade sceptre, which is kept by the dukes, is called *xin gui* 信圭. A seven Cun 寸 long jade sceptre, which is kept by the counts, is called *gong gui* 躬圭. The jade device *mao* 冒, carried by the son of heaven, is four Cun 寸 (7.9 cm) long. He uses it when he receives the princes during an audience. The son of heaven uses jade of a pure colour. The princes use jade of mixed colours. The dukes use impure jade, and the counts a jade of which one half contains other minerals. The descendants of the princes use ritual objects made from leather and silk. A silk ribbon is wound around the middle of the jade sceptre of the son of heaven. The jade sceptres that are fixed

7 判规，判可解为半，规解为圆。

8 "璧羡度尺"，璧指玉璧。郑玄认为"羡犹延，其袤一尺而广狭焉。""好"指的是璧中央的孔。《尔雅》有"肉倍好谓之璧"之语，"肉"指孔外璧的实体。既然此处"好三寸"，则知每侧璧宽亦为三寸，璧总体宽共九寸。

9 瑑，指玉器上的饰纹。瑑圭璋，指有瑑饰的圭璋。

* See commentary: Jade.

onto the four walls of the palace room are one Chi 尺, two Cun 寸 (23.6 cm) long. They are used for praying to heaven. The great sceptre *da gui* 大圭 is three Chi 尺 (59 cm) long. On the oblong shaft on the top is a hammer-shaped part. The son of heaven carries it on his robes. The earth sceptre *tu gui* 土圭 is one Chi 尺, five Cun 寸 (29.5 cm) long and is used to measure shadow length and to survey the earth. The jade spoon *guan gui* 祼圭 is one Chi 尺, two Cun 寸 (23.6 cm) long and is used to make wine sacrifices in the temple. The sceptre *wan gui* 琬圭 is nine Cun (17.7 cm) long and comes with a wooden coaster. It symbolises the virtue of the ruler. The sceptre *yan gui* 琰圭 is nine Cun 寸 (17.7 cm) long and is shaped like a semicircular ring. It is used to keep evil away and to improve the deeds of the princes. The diameter of the jade disc *bi xian* 璧羡 is one Chi 尺 (19.7 cm), and the diameter of the hole three Cun 寸 (5.9 cm). It serves as a standard for length. The jade sceptre *gui* 圭 and the jade disc *bi* 璧, with a size of five Cun 寸 (9.8 cm), are used to make sacrifices to the sun, the moon and the stars. The vassal princes use jade discs *bi* 璧 and jade tubes *cong* 琮 as presents for the son of heaven. The jade sceptre *gu gui* 谷圭 is seven Cun 寸 (13.8 cm) long. The son of heaven uses it for the marriage of his daughters.

The jade sceptres *da zhang* 大璋 and *zhong zhang* 中璋 are nine Cun 寸 (17.7 cm) long, and the jade sceptre *bian zhang* 边璋 is seven Cun 寸 (13.8 cm) long. It tapers to four Cun 寸 (7.9 cm) and is one Cun 寸 (2.0 cm) thick. A sacrificial spoon from gold is set with turquoises on the outside; the inside of the spoon is lacquered in red, and the outflow opening for the sacrificial wine is one Cun 寸 (2.0 cm), whereas the spoon part has a diameter of four Cun 寸 (7.9 cm). A wooden coaster belongs to it. The son of heaven uses it when hunting, to direct prayers to the mountains and rivers. Before he sacrifices a horse to the gods of the mountains and rivers, he sacrifices wine together with the official responsible for sacrifices. The sceptre *da zhang* 大璋 also is seven Cun 寸 (13.8 cm) long. The vassal princes use it for the marriage of their daughters. The sceptres *zhuan gui* 瑑圭 and *zhuan zhang* 瑑璋, carved with raised patterns, are eight Cun 寸 (15.7 cm) long. The jade tube *bi cong* 璧琮 is eight Cun 寸 long. These jade devices serve the vassal princes as presents for visits. The sceptres *ya zhang* 牙璋[10] and *zhong zhang* 中璋 are seven Cun 寸 (13.8 cm) long; they taper to two Cun 寸 (3.9 cm) and are one Cun 寸 (2.0 cm) thick. They are used to give the army marching orders and to deploy parts of the army. The jade tube *zu cong* 駔琮 is five Cun 寸 (9.8 cm) long. The queen uses it as a weight.

10 *Sceptre ya zhang* 牙璋 – the sceptre *ya zhang* 牙璋 was divided into two parts. The two parts were handed over to two partners of an order or contract. When the two halves could be put together, this served to prove the authenticity of the delivered document.

The jade tube *da cong* 大琮 is twelve Cun 寸 (23.6 cm) long; it tapers to four Cun 寸 (7.9 cm) and is one Cun 寸 (2.0 cm) thick. It is also called *nei zhen* 內鎮 and is kept by the queen. The jade tube *zu cong* 駔琮 has a diameter of seven Cun 寸 (13.8 cm); its hole is one and a half Cun 寸 (3 cm). It serves the king as a weight. Two five-Cun 寸 (9.8 cm) long sceptres *gui* 圭 lay with their base planes smoothly on top of each other. They are used to make sacrifices to the earth god and the gods of the four directions. The eight-Cun 寸 (15.7 cm) long raised decorated jade tubes *zhuan cong* 瑑琮 serve the vassal princes as presents for the spouse of the country's ruler. A small table, decorated with jade, has a height of twelve Cun 寸 (23.6 cm); twelve dates and twelve chestnuts are offered on it. For the vassal prince, nine fruits are offered, and for the dignitaries, five fruits each. In this way, the spouse of the ruler expresses her gratitude for their services. The base plane of the sceptre *zhang* 璋 tapers and does not carry decorations. It is used for making sacrifices to the gods of the mountains and the rivers, and to pay the officials in kind.

Teil 2: 18. Jade*

Die Arbeit des Jadeschleifers. Das Jadeszepter *zhen gui* 鎮圭, das dem Himmelssohn vorbehalten ist, ist ein Chi 尺 zwei Cun 寸 (23,6 cm) lang. Ein neun Cun 寸 (17,7 cm) langes Jadeszepter, das von den Fürsten verwahrt wird, heißt *huan gui* 桓圭. Ein sieben Cun 寸 (13,8 cm) langes Jadeszepter, das von den Herzögen verwahrt wird, heißt *xin gui* 信圭. Ein sieben Cun 寸 langes Jadeszepter, das von den Grafen verwahrt wird, heißt *gong gui* 躬圭. Das vom Himmelssohn getragene Jadegerät *mao* 冒 ist vier Cun 寸 (7,9 cm) lang. Er verwendet es, wenn er die Fürsten in Audienz empfängt. Der Himmelssohn benutzt reinfarbene Jade. Die Fürsten benutzen gemischt farbige Jade. Die Herzöge benutzen unreine Jade, und die Grafen eine Jade, die zur Hälfte andere Mineralien enthält. Die Nachkommen der Fürsten benutzen Ritualgegenstände aus Leder und Seide. In der Mitte des Jadeszepters des Himmelssohns ist ein Seidenband gebunden. Die an den vier Wänden des Palastraums angebrachten Jadeszepter sind ein Chi 尺 zwei Cun 寸 (23,6 cm) lang. Sie dienen dazu, zum Himmel zu beten. Das große Szepter *da gui* 大圭 ist drei Chi 尺 (59,1 cm) lang. Auf dem länglichen Schaft befindet sich oben ein hammerförmiges Teil. Der Himmelssohn trägt es an seinem Gewand. Das Erdszepter *tu gui* 土圭 ist ein Chi 尺 fünf Cun 寸 (29,5 cm) lang und dient zur

* Siehe Kommentar: Jade.

Messung der Schattenlänge und zur Vermessung der Erde. Der Jadelöffel *guan gui* 祼圭 ist ein Chi 尺 zwei Cun 寸 (23,6 cm) lang und dient für Weinopfer im Tempel. Das Szepter *wan gui* 琬圭 ist neun Cun 寸 (17,7 cm) lang, und zu ihm gehört ein hölzerner Untersatz. Es symbolisiert die Tugend des Herrschers. Das Szepter *yan gui* 琰圭 ist neun Cun 寸 (17,7 cm) lang und hat die Form eines halbkreisförmigen Rings. Es dient dazu, das Böse fernzuhalten und das Handeln der Fürsten zu verbessern. Der Durchmesser der Jadescheibe *bi xian* 璧羨 beträgt ein Chi 尺 (19,7 cm) und der Durchmesser der Bohrung drei Cun 寸 (5,9 cm). Es dient als Längennormal. Jadeszepter *gui* 圭 und Jadescheiben *bi* 璧 mit einem Maß von fünf Cun 寸 (9,8 cm) dienen für Opfer an die Sonne, den Mond und die Sterne. Jadescheiben *bi* 璧 und Jaderöhren *cong* 琮 benutzen die Lehnsfürsten als Geschenk für den Himmelssohn. Das Jadeszepter *gu gui* 谷圭 ist sieben Cun 寸 (13,8 cm) lang. Der Himmelssohn benutzt es für die Verheiratung seiner Töchter.

Die Jadeszepter *da zhang* 大璋 und *zhong zhang* 中璋 sind neun Cun 寸 (17,7 cm) lang, und das Jadeszepter *bian zhang* 邊璋 ist sieben Cun 寸 (13,8 cm) lang. Es verjüngt sich auf vier Cun 寸 (7,9 cm) und ist ein Cun 寸 dick (2,0 cm). Ein Opferlöffel aus Gold ist außen mit Türkisen besetzt, die Innenseite des Löffels ist rot lackiert, und die Austrittsöffnung für den Opferwein ist ein Cun 寸 (2,0 cm) groß, während der Löffelteil einen Durchmesser von vier Cun 寸 (7,9 cm) hat. Dazu gehört ein hölzerner Untersatz. Der Himmelssohn benutzt ihn bei der Jagd, um an die Berge und Flüsse Gebete zu richten. Bevor er den Göttern der Berge und Flüsse ein Pferd opfert, opfert er mit dem für das Opfer zuständigen Beamten Wein. Das Szepter *da zhang* 大璋 ist auch sieben Cun 寸 (13,8 cm) lang. Die Lehnsfürsten benutzen es für die Verheiratung ihrer Töchter. Die mit erhabenen Mustern geschnitzten Szepter *zhuan gui* 瑑圭 und *zhuan zhang* 瑑璋 sind acht Cun 寸 (15,7 cm) lang. Die Jaderöhre *bi cong* 璧琮 ist acht Cun 寸 lang. Diese Jadegeräte dienen den Lehnsfürsten als Besuchsgeschenk. Die Szepter *ya zhang* 牙璋[11] und *zhong zhang* 中璋 sind sieben Cun 寸 (13,8 cm) lang, verjüngen sich auf zwei Cun 寸 (3,9 cm) und sind ein Cun 寸 (2,0 cm) dick. Sie dienen dazu, dem Heer Marschbefehle zu erteilen, und Heeresteile zu versetzen. Die Jaderöhre *zu cong* 駔琮 ist fünf Cun 寸 (9,8 cm) lang. Die Königin benutzt sie als Gewichtsstück. Die Jaderöhre *da cong* 大琮 ist zwölf Cun 寸 (23,6 cm) lang, verjüngt sich auf vier Cun 寸 (7,9 cm) und ist ein Cun 寸 (2 cm) dick. Sie heißt auch *nei zhen* 内鎮 und wird von der

11 *Szepter ya zhang* 牙璋 – Das Szepter *ya zhang* 牙璋 war in zwei Teile geteilt. Die beiden Teile wurden zwei Partnern eines Befehls oder Vertrags übergeben. Wenn die beiden Hälften zusammengelegt werden konnten, diente dies dazu, die Authentizität eines beförderten Dokuments zu beweisen.

Königin verwahrt. Die Jaderöhre *zu cong* 駔琮 hat einen Durchmesser von sieben Cun 寸 (13,8 cm), ihre Bohrung ist anderthalb Cun 寸 (3 cm) groß. Sie dient dem König als Gewichtsstück. Zwei fünf Cun 寸 (9,8 cm) lange Szepter *gui* 圭 liegen mit ihren Grundflächen glatt aufeinander. Sie dienen für die Opfer an den Erdgott und die Götter der vier Himmelsrichtungen. Die acht Cun 寸 (15,7 cm) langen, erhaben verzierten Jaderöhren *zhuan cong* 瑑琮 dienen den Lehnsfürsten als Geschenk für die Gattin des Landesherrschers. Ein mit Jade verziertes Tischchen hat eine Höhe von zwölf Cun 寸 (23,6 cm), darauf werden je zwölf Datteln und Kastanien dargeboten. Für die Lehnsfürsten werden je neun Früchte und für die Würdenträger je fünf Früchte dargeboten. Damit bedankt sich die Gattin des Herrschers für ihre Dienste. Die Grundfläche des Szepters *zhang* 璋 verjüngt sich und trägt keine Verzierungen. Es dient für Opfer an die Götter der Berge und Flüsse und um Beamtengehälter in Naturalien auszuzahlen.

19AC. 栉人（阙）

19MC. 栉人（已阙）

19. Comb Maker (Missing)

19. Der Kammmacher (fehlt)

20AC. 雕人（阙）

20MC. 雕人（已阙）

20. Bone Carver (Missing)

20. Der Knochenschnitzer (fehlt)

21AC. 磬氏

磬氏为磬，倨句一矩有半，其博为一，股为二，鼓为三。参分其股博，去一以为鼓博。参分其鼓博，以其一为之厚。已上，则摩其旁；已下，则摩其耑。

21MC. 磬氏

磬氏制作磬，要使其两个上边的开阖程度达到一个磬折的大小，如果博宽为一，那么股长就为二，鼓长则为三。鼓博宽度是股博宽度的三分之二，磬的厚度是鼓博宽度的三分之一。制成的磬如果音调太高，就研磨磬的侧面；音调低了，则研磨磬的端面。

21. The Manufacturer of Chimes*

The official Qing Shi 磬氏 manufactures stone chimes. The top angle is 135°. If one takes the width of the side *gu* 股 (thigh) as a unit of length, then the length of the side *gu* 股 amounts to two length units, and the length of the side *gu* 鼓 (drum) three length units. The width of the side *gu* 鼓 (drum) amounts to two thirds of the width *gu* 股 (thigh). One third of the width of *gu* 鼓 (drum) provides the thickness of the chime. If the sound of the chime is too high, then a bit of the two flat sides *pang* 旁 is ground off. If the sound is too deep, then one grinds a bit off of the lower front side *duan* 端.

21. Der Hersteller von Klangsteinen**

Der Beamte Qing Shi 磬氏 fertigt Klangsteine an. Der obere Scheitelwinkel beträgt 135°. Nimmt man die Breite des Schenkels *gu* 股 als eine Längeneinheit, dann beträgt die Länge von *gu* 股 (Schenkel) zwei Längeneinheiten und die Länge von *gu* 鼓 (Trommel) drei Längeneinheiten. Die Breite von *gu* 鼓 (Trommel) beträgt zwei Drittel der Breite von *gu* 股 (Schenkel). Ein Drittel der Breite von *gu* 鼓 (Trommel) ergibt die Dicke des Klangsteins. Wenn der Klang des Klangsteins zu hoch ist, dann schleift man an die beiden flachen Seiten *pang* 旁. Ist der Klang zu tief, dann schleift man an der unteren Stirnseite *duan* 端.

22AC. 矢人

矢人为矢。鍭矢参分，茀矢参分，一在前，二在后。兵矢、田矢五分，二在前，三在后。杀矢七分，三在前，四在后。参分其长，而杀其一。五分

* See commentary: Musical Instruments.

** Siehe Kommentar: Musical Instruments.

其长，而羽其一。以其笴厚为之羽深。水之，以辨其阴阳。夹其阴阳，以设其比。夹其比，以设其羽。参分其羽，以设其刃，则虽有疾风，亦弗之能惮矣。刃长寸围寸，铤十之，重三垸。前弱则俯，后弱则翔，中弱则纡，中强则扬。羽丰则迟，羽杀则趮。是故夹而摇之，以眡其丰杀之节也。桡之，以眡其鸿杀之称也。凡相笴，欲生而抟。同抟，欲重；同重，节欲疏；同疏，欲栗。

22MC. 矢人

矢人为矢。鍭矢参分，茀矢参分，一在前，二在后。兵矢、田矢五分，二在前，三在后。杀矢七分，三在前，四在后。参分其长，而杀其一。五分其长，而羽其一。以其笴厚为之羽深。水之，以辨其阴阳。夹其阴阳，以设其比。夹其比，以设其羽。参分其羽，以设其刃，则虽有疾风，亦弗之能惮矣。刃长寸围寸，铤十之，重三垸。前弱则俯，后弱则翔，中弱则纡，中强则扬。羽丰则迟，羽杀则趮。是故夹而摇之，以眡其丰杀之节也。桡之，以眡其鸿杀之称也。凡相笴，欲生而抟。同抟，欲重；同重，节欲疏；同疏，欲栗。

22MC. 矢人

矢人负责制箭。鍭矢和茀矢，箭的前三分之一部分的重量与后三分之二部分的重量相等。兵矢和田矢，箭的前五分之二部分的重量与后五分之三部分的重量相等。杀矢，箭的前七分之三的重量与后七分之四部分的重量相等。[12] 把箭杆[13] 长度分为三份，从前面三分之一处开始向前逐渐削细。把箭杆长度分为五份，后面的五分之一处安装箭羽。根据箭杆的粗细设定箭羽安装的深浅。将箭杆浮在水上，以辨别箭杆木材的阴面和阳面。向上为阴，向下为阳。在垂直平分阴面和阳面的地方设置箭括（用于扣弦之凹槽）。在平分箭括的地方安装箭羽。箭镞长度是箭羽长度的三分之一，这样即使有强风，也不会对箭的飞行有多大影响。箭镞长一寸，它的周长也是一寸，那么铤（箭镞下端没入箭杆的部分）长为一尺，箭镞重三垸。如果箭杆前半部分柔弱，箭飞行过程中就容易下栽；后半部分柔弱，箭飞行过程中就容易上飘；中间部分柔弱，箭飞行过程中容易偏斜；中间部分过于刚硬，箭的飞行会上扬。如果箭羽植得过密，箭的飞行速度会变慢；箭

12 鍭矢、茀矢、兵矢、田矢、杀矢等，是不同用途的几种箭。鍭矢箭镞重，用于近射，杀伤力较大。茀矢箭镞较田矢稍轻，用途与田矢相仿，用于打猎弋射。兵矢利于火射，用于守城、车战。杀矢箭镞亦重，用途与鍭矢相仿。

13 笴，箭杆。

羽稀疏，箭飞行中容易左右摇晃。因此要用手指夹住箭杆摇动它，以观察箭羽的密疏是否合适。弯曲箭杆，以观察其粗细强弱是否得当。凡是选择用作箭杆的竹材，希望它天生浑圆；同样是浑圆的，要选那些重实的；同样是重实的，要选竹节间距大的；同样是竹节间距大的，要选那些颜色如栗的。

22. The Arrow Tip Manufacturer*

The arrow tip manufacturer manufactures arrow tips. For the arrow tips *hou shi* 鍭矢 and *bo shi* 茀矢, one third of the front part weighs as much as two thirds of the back part. For the army arrows and hunting arrows, two fifths of the front part weigh as much as three fifths of the back part. For the deadly arrows, three sevenths of the front part weigh as much as four sevenths of the back part. One third of the front part tapers until its diameter corresponds to the diameter of the arrow tip. Onto one fifth of the back part of the arrow shaft, arrow feathers are fixed. The depth at which the feathers are put into the arrow shaft equals the radius of the arrow shaft. If one puts the arrow into water, one finds that the Yin wood rises, whereas the Yang wood sinks.[14] Perpendicular to the layers of Yin and Yang wood at the arrow's end, a notch for the bow string is made. In this notch, the arrow feathers are fixed. The length of the arrow tip amounts to one third of the feather length, so that the arrow will not deviate from its path, even with a strong wind. The arrow tip has a length of one Cun 寸 (2.0 cm); its circumference is one Cun 寸, the part of the arrow shaft that the tip is connected to is one Chi 尺 (19.7 cm) long, and the arrow weighs three Huan 锾 (17.9 g). If the front part of the arrow shaft is weak, the arrow flies lower than usual. If the back part of the arrow shaft is weak, the arrow flies higher than usual. If the middle part of the arrow shaft is weak, it flies on a wavy line. If the middle part of the arrow shaft is firm, it flies on a straight line. If too many feathers are placed on the arrow shaft, they slow down the speed. If the arrow has too few or no feathers, the arrow flies crooked. Therefore, one swings the arrow between one's fingers to test whether the number of feathers is suitable. One bends the arrow shaft to test whether the arrow shaft is sufficiently firm. When one chooses the material for the arrow shaft, it is desirable that it is grown straight and round. Besides possessing roundness, the arrow

---

* See commentary: Bronze objects.

14 *One finds that the Yin 阴 wood rises, whereas the Yang 阳 wood sinks* – For bamboo, it is known that the wood on the Yang side is denser and firm, and on the Yin side less dense and firm.

must be sufficiently heavy. Besides its weight, the distance between the bamboo knots should be large. Besides the large distance of the knots, its colour should be chestnut brown.

22. Der Pfeilspitzenmacher*

Der Pfeilspitzenmacher stellt Pfeilspitzen her. Bei den Pfeilspitzen *hou shi* 鍭矢 und bo *shi* 茀矢 wiegen ein Drittel des vorderen Teils so viel wie zwei Drittel des hinteren Teils. Bei den Heerespfeilen und den Jagdpfeilen wiegen zwei Fünftel des vorderen Teils so viel wie drei Fünftel des hinteren Teils. Bei den tödlichen Pfeilen wiegen drei Siebentel des vorderen Teils so viel wie vier Siebentel des hinteren Teils. Ein Drittel des vorderen Teils des Pfeilschafts verjüngt sich, bis sein Durchmesser dem Durchmesser der Pfeilspitze entspricht. Auf einem Fünftel des hinteren Teils des Pfeilschafts werden die Pfeilfedern befestigt. Die Tiefe, mit der die Federn in den Pfeilschaft gesteckt werden, ist gleich dem Radius des Pfeilschafts. Wenn man die Pfeile in Wasser legt, erkennt man, dass das Yin 阴-Holz aufsteigt, während das Yang 阳-Holz sinkt.[15] Senkrecht zur Schichtung von Yin 阴- und Yang 阳-Holz wird am Pfeilende die Einkerbung für die Bogensehne eingearbeitet. In dieser Einkerbung werden die Pfeilfedern befestigt. Die Länge der Pfeilspitze beträgt ein Drittel der Federlänge, so dass der Pfeil auch bei starkem Wind nicht von seiner Bahn abweicht. Die Pfeilspitze hat eine Länge von einem Cun 寸 (2,0 cm), ihr Umfang beträgt ein Cun 寸, der Teil des Pfeilschafts, mit dem die Spitze verbunden ist, ist ein Chi 尺 (19,7 cm) lang, und der Pfeil wiegt drei Huan 锾 (17,9 g). Wenn der vordere Teil des Pfeilschafts schwach ist, dann fliegt der Pfeil niedriger als normal. Wenn der hintere Teil des Pfeilschafts schwach ist, fliegt der Pfeil höher als normal. Wenn der mittlere Teil schwach ist, fliegt er eine Schlangenlinie. Wenn der mittlere Teil des Pfeilschafts fest ist, wird er auf gerader Linie fliegen. Wenn auf dem Pfeilschaft zu viele Federn sitzen, verlangsamen sie die Schnelligkeit. Wenn der Pfeil zu wenige oder gar keine Federn hat, fliegt der Pfeil schief. Darum schwenkt man den Pfeil mit den Fingern hin und her, um zu prüfen, ob die Anzahl der Federn geeignet ist. Man biegt den Pfeilschaft, um zu prüfen, ob der Pfeilschaft ausreichend fest ist. Wenn man das Material für den Pfeilschaft auswählt, möchte man, dass es gerade gewachsen und rund

---

* Siehe Kommentar: Bronze Objects.

15 *Wenn man die Pfeile in Wasser legt, erkennt man, dass das Yin 阴-Holz aufsteigt, während das Yang 阳-Holz sinkt* – Von Bambus ist bekannt, dass das Holz auf der Yang 阳-Seite dichter und fester und auf der Yin 阴-Seite weniger dicht und fest ist.

ist. Neben der Rundheit muss der Pfeil ausreichend schwer sein. Neben seinem Gewicht soll der Abstand zwischen den Bambusknoten groß sein. Neben dem großen Abstand der Knoten soll seine Farbe kastanienbraun sein.

23AC. 陶人

陶人为甗，实二鬴，厚半寸，唇寸。盆实二鬴，厚半寸，唇寸。甑实二鬴，厚半寸，唇寸，七穿。鬲实五觳，厚半寸，唇寸。庾实二觳，厚半寸，唇寸。

旊人为簋，实一觳，崇尺，厚半寸，唇寸。豆实三而成觳，崇尺。凡陶旊之事，髻垦薜暴不入市。器中膊，豆中县。膊崇四尺，方四寸。

23MC. 陶人

陶工制作甗（蒸具，分上下两层。上层若甑，可以蒸；下层若鬲，可以煮）、盆、甑、鬲和庾（一种瓦器）。甗的容积是二鬴，壁厚半寸，唇宽一寸；盆的容积二鬴，壁厚半寸，唇宽一寸；甑的容积是二鬴，壁厚半寸，唇宽一寸，底部穿有七个小孔；鬲的容积是五觳，壁厚半寸，唇宽一寸；庾的容积是二觳，壁厚半寸，唇宽一寸。

23. The Potter*

The official Taoren 陶人 manufactures the cooking vessel *yan* 甗. Its volume is two Fu 鬴 (25.6 l). The wall is half a Cun 寸 (1.0 cm) thick, and one Cun 寸 (2.0 cm) on the top rim. The volume of the basin *pen* 盆 is two Fu 鬴; its wall is half a Cun 寸 thick and one Cun 寸 at the rim. The steaming vessel *zeng* 甑 has a volume of two Fu 鬴; its wall is half a Cun 寸 thick and the top rim one Cun 寸. On the bottom, it has seven small holes. The three-legged vessel *li* 鬲 has a volume of five Hu 觳 (6.0 l); its wall is half a Cun 寸 and its top rim one Cun 寸 thick. The vessel *yu* 庾 has a volume of two Hu 觳 (2.4 l). Its wall is half a Cun 寸 and its top rim one Cun 寸 thick.

The official Fangren 旊人 manufactures the vessels *gui* 簋 to store food. Their volume is one Hu 觳 (1.2 l); they are one Chi 尺 (19.7 cm) high, their wall is half a Cun 寸 thick and the rim one Cun 寸. The volume of the vessel *dou* 豆

* See commentary: Ceramics and Porcelain.

is one third of a Hu 觳 (0.4 l) and is one Chi 尺 (19.7 cm) high. Generally, vessels from ceramics and burnt clay that are crooked, chipped, torn and uneven are not allowed to be brought to the market of the officials. The ceramic vessels must correspond to the height gauge. The grips of the vessel *dou* 豆 must be fastened vertically, so that one can hang it with a rope. The form gauges are four Chi 尺 (78.6 cm) high and have a square cross section with an edge length of four Cun 寸 (7.9 cm).

23. Der Töpfer*

Der Beamte Taoren 陶人 stellt das Kochgefäß *yan* 甗 her. Sein Volumen beträgt zwei Fu 鬴 (25,6 l). Die Wand ist einen halben Cun 寸 (1,0 cm) und am oberen Rand ein Cun 寸 (2,0 cm) dick. Das Volumen des Beckens *pen* 盆 beträgt zwei Fu 鬴, seine Wand ist einen halben Cun 寸 und der Rand ein Cun 寸 dick. Das Dämpfgefäß *zeng* 甑 hat ein Volumen von zwei Fu 鬴, seine Wand ist einen halben Cun 寸 und der obere Rand ein Cun 寸 dick. Am Boden hat es sieben kleine Löcher. Das dreifüßige Gefäß *li* 鬲 hat ein Volumen von fünf Hu 觳 (6,0 l), seine Wand ist einen halben Cun 寸 und sein oberer Rand ein Cun 寸 dick. Das Gefäß *yu* 庾 hat ein Volumen von zwei Hu 觳 (2,4 l). Seine Wand ist einen halben Cun 寸 und sein oberer Rand ein Cun 寸 dick.

Der Beamte Fangren 旊人 stellt Gefäße *gui* 簋 zum Aufbewahren von Nahrungsmitteln her. Ihr Volumen beträgt ein Hu 觳 (1,2 l), sie sind ein Chi 尺 (19,7 cm) hoch, ihre Wand ist einen halben Cun 寸 und der Rand ein Cun 寸 dick. Das Volumen des Gefäßes *dou* 豆 beträgt 1/3 Hu 觳 (0,4 l) und ist ein Chi 尺 (19,7 cm) hoch. Allgemein dürfen Gefäße aus Keramik und gebranntem Lehm, die schief, angeschlagen, gerissen und uneben sind, nicht auf den Markt für die Beamten gebracht werden. Die Keramikgefäße müssen der Höhenlehre entsprechen. Die Griffe des Gefäßes *dou* 豆 müssen senkrecht angebracht sein, so dass man es mit einem Seil aufhängen kann. Die Formlehren sind vier Chi 尺 (78,6 cm) hoch und haben einen quadratischen Querschnitt mit vier Cun 寸 (7,9 cm) Kantenlänge.

24AC. 梓人

梓人为筍虡。天下之大兽五:脂者、膏者、赢者、羽者、鳞者。宗庙之事，脂者膏者以为牲；赢者、羽者、鳞者以为筍虡。外骨、内骨，却

* Siehe Kommentar: Ceramics and Porcelain.

行、仄行、连行、纡行、以脰鸣者、以注鸣者、以旁鸣者、以翼鸣者、以股鸣者、以胸鸣者，谓之小虫之属，以为雕琢。厚唇弇口，出目短耳，大胸燿后，大体短脰，若是者谓之赢属。恒有力而不能走，其声大而宏。有力而不能走，则于任重宜；大声而宏，则于钟宜，若是者以为钟虡，是故击其所县，而由其虡鸣。锐喙决吻，数目顅脰，小体骞腹，若是者谓之羽属。恒无力而轻，其声清扬而远闻。无力而轻，则于任轻宜；其声清扬而远闻，于磬宜，若是者以为磬虡，故击其所县而由其虡鸣。小首而长，抟身而鸿，若是者谓之鳞属，以为筍。凡攫閷援簭之类，必深其爪，出其目，作其鳞之而。深其爪，出其目，作其鳞之而，则于眡必拨尔而怒。苟拨尔而怒，则于任重宜，且其匪色必似鸣矣。爪不深，目不出，鳞之而不作，则必穨尔如委矣。苟穨尔如委，则加任焉，则必如将废措，其匪色必似不鸣矣。

梓人为饮器，勺一升，爵一升，觚三升。献以爵而酬以觚，一献而三酬，则一豆矣；食一豆肉，饮一豆酒，中人之食也。凡试梓饮器，乡衡而实不尽，梓师罪之。

梓人为侯，广与崇方。参分其广，而鹄居一焉。上两个，与其身三；下两个，半之。上纲与下纲出舌寻，縜寸焉。张皮侯而栖鹄，则春以功；张五采之侯，则远国属；张兽侯，则王以息燕。祭侯之礼，以酒脯醢，其辞曰:“惟若宁侯，毋或若女不宁侯，不属于王所，故抗而射女。强饮强食，诒女曾孙诸侯百福。”

24MC. 梓人

梓人仿照动物造型制作筍虡。[16] 天下的动物大致可分为五类:脂类、膏类、赢类、羽毛类、鳞类。[17] 脂类膏类动物一般用于宗庙祭祀的牺牲，赢类、羽毛类、鳞介类可以作为制作筍虡时的仿照和纹饰。骨在体外的、在体内的，倒着行的、侧着旁行的、连贯而行的、屈伸而行的以及用颈项发声的、用尖尖的嘴发声的、用身体的旁侧发声的、用翅膀发声的、用大腿发声的和用胸发声的，等等，这些称为小虫之类。它们可以用作筍虡上的纹饰。赢类动物唇厚口深，眼睛凸起，耳朵短小，胸大臀瘦，体长脖短，体健有力，不善奔跑，声音洪亮。有力而不善于奔跑，则适合于承重；声音洪亮，则适合于钟。由此，钟虡的造型适合以这类动物为参照，这样当敲击其所悬挂的钟时，其洪亮的声音好像由钟虡本身发出

16 筍虡，古代悬挂钟、磬等乐器的架子。立柱为虡，横木为筍。因筍虡需有一定的造型，古人取法于动物，故这一段体现了古人对动物的一些认识。

17 脂者应指有角的动物。膏者指无角的动物。赢指裸身的人，这里是把自然界的人类作为动物的一种。羽者指鸟类。鳞者，按《考工记》的描述，应指龙。

来的一样。羽毛类动物嘴尖而开，眼小颈长，身体瘦小，腹部收敛，力气小，体重轻，发声清脆、传播得远。力小体轻，只能承受较轻的负荷。因为声音轻扬，传播得远，适合作磬虡，这样当敲击其所悬挂的磬时，其清脆的声音好像由磬虡发出来的一样。鳞类动物头小身长，身体圆而均匀，适宜于作筍的造型。凡是善于扑杀抓咬的动物，一定是深藏其爪，突出其目，鳞羽振起，看上去一副勃然大怒的样子。如果是勃然大怒的样子，就适合于承重，并且其神态一定也像是在鸣叫的样子。如果爪没有收藏、眼睛也没有突出，鳞羽没有振起，那就一定是一副萎靡不振的样子。如果是萎靡不振的样子，让其承重，看上去就会有不堪重负的感觉，其神采也不像是要鸣叫的样子。

24. The Carpenter Ziren*

The carpenter Ziren 梓人 manufactures stands for bell chimes. (The stands are decorated with animal figures.) The large animals under the sky form five groups:[18] Animals with long horns, animals without long horns, naked men, feathered animals, scaly animals. Animals with and without long horns serve in the temples as sacrificial animals.

Naked men, feathered and scaly animals serve as figures for the columns or as a cross-beam of the stand. Invertebrates and vertebrates can go backwards, sideways, one behind the other, and in a winding way. The animals that emit sounds with their neck, beak, abdominal wall, wings, legs or chest are called small reptiles. They are found in the stands as carved decoration. The representation of humans is distinguished by the following features: thick lips, a closed mouth, protruding eyes, small ears, a wide chest, a narrow back, a big body and a short neck. The figures appear as if they would stride along powerfully, as if their voice would be tremendous. The powerful appearance matches with the carrying of the load; their tremendous voice matches with the bells. Such figures are used for the columns of the bell chime stand. If one rings the hung bells, the columns seem to resonate as well.

The animals that have a pointed beak, an opened mouth, small eyes, a long neck, a small body and a flat abdomen are called feathered animals. They appear quick and weak. Their voice is pure and penetrating. Their quick and

* See commentary: Musical Instruments.

18 *The large animals under the sky form five groups* – This paragraph on the stands of chime sets contains a classification of living beings. It is of interest that naked humans also belong to the fauna. However, in excavated stands of chime sets the humans carrying the cross-beams are dressed and carry a sword.

weak appearance and their penetrating voice matches with the stone chimes. Therefore, these animals serve as columns for stone chimes. If one rings the hung stone chimes, the columns of the stand seem to resonate as well. Animals with small, oblong heads and evenly round bodies are called scaly animals. They are found as figures in the cross-beams.

In the presentation of wild beasts, one must carve out the claws, and the eyes must protrude. Scales or hair must stand up straight so that they look as if they suddenly flew into a rage. This hot-tempered expression is underlined by the weight of the figure, and it should look as if it would roar strongly. If the claws are not carved out, the eyes do not protrude and the scales or the hair do not stand up straight, then the figure seems to be weak and depressed. If it seems to be weak and depressed, then one gets the impression that the figure cannot carry the load. In this case, the figure seems to be exhausted, and its expression makes it seem like it does not roar.

The carpenter Ziren 梓人 manufactures drinking vessels. The volume of ladles is one Sheng 升 (0.2 l). The three-legged drinking vessels *jue* 爵 have a volume of one Sheng 升, and the drinking vessels *gu* 觚 a volume of three Sheng 升 (0.6 l). Wine is presented to the guest with the drinking vessel *jue* 爵, whereas the drinking vessel *gu* 觚 is used to propose a toast. One presents one Sheng 升 (0.2 l) wine and proposes a toast with three Sheng 升 (0.6 l), which together yields one Dou 豆 (0.8 l). One eats one Dou 豆 meat and drinks one Dou 豆 wine; that is the amount of food of ordinary men. Generally, the drinking vessels manufactured by the carpenter Ziren 梓人 are tested as follows: if the vessel is raised horizontally to the mouth and wine remains in it, the superintendent of the carpenter Ziren 梓人 must punish him.

The carpenter Ziren 梓人 manufactures targets. The width and height of the target together form a square. The field in the middle takes up one third of the target's width. The two fixtures on the top, on the right and left, have the same width as the target, so that the total width amounts to threefold the width of the actual target. The two fixtures at the bottom right and left are half the width of the fixtures on top. The ropes on the fixtures on top and at the bottom are eight Chi 尺 (157.6 cm) long each. They are fixed on a knob with a one Cun 寸 (2.0 cm) diameter. If one attaches a target made from leather, the middle field is sewed onto the middle of the target. In spring, a big archer ritual is held to compare the abilities of the vassal princes. For this, a target is used whose fields are painted with five colours. The vassal princes from the far states gather at the court. A target painted with animals is spread out. While the king feeds the vassal princes, the archers' ritual is held. During this ritual, wine, dried meat and meat sauce are sacrificed. The sacrificial prayer runs: "I only take obedient and meritorious vassal princes as an example. Under no circumstances should we be confused by those disobedient vassal princes who do not gather in the king's court and do not follow the alliance. Therefore, I will resist them and shoot them down with the bow. The obedient vassal princes should receive plenty of drinks and food. May great luck be passed to your sons and grandsons so that you, generation by generation, are my vassal princes!"

24. Der Tischler Ziren*

Der Tischler Ziren 梓人 stellt die Gestelle für Glockenspiele her. (Die Gestelle sind mit Tierfiguren geschmückt.) Die großen Tiere unter dem Himmel bilden fünf Gruppen:[19] Tiere mit langen Hörnern, Tiere ohne lange Hörner, nackte Menschen, gefiederte Tiere, schuppige Tiere. Tiere mit und ohne lange Hörner dienen in den Tempeln als Opfertiere. Nackte Menschen, gefiederte und schuppige Tiere dienen als Figuren für die Säulen oder das Querholz des Gestells. Panzertiere und Wirbeltiere können rückwärts, seitwärts, hintereinander und sich schlängelnd gehen. Die Tiere, die mit dem Hals, dem Schnabel, mit der Bauchwand, mit den Flügeln, den Beinen oder mit der Brust Laute von sich geben, heißen kleine Kriechtiere. Sie finden sich in den Gestellen als geschnitzte Verzierung. Die Darstellung von Menschen zeichnet sich durch folgende Formen aus: Dicke Lippen, geschlossener Mund, hervorstehende Augen, kleine Ohren, breite Brust, schmaler Rücken, großer Körper und kurzer Hals. Die Figuren erscheinen, als würden sie machtvoll einherschreiten, als wäre ihre Stimme gewaltig. Die machtvolle Erscheinung passt zum Tragen der Last, ihre gewaltige Stimme passt zu den Glocken. Solche Figuren werden für die Säulen des Glockenspielgestells verwendet. Wenn man die aufgehängten Glocken anschlägt, scheinen die Säulen auch mitzuklingen.

Die Tiere, die einen spitzen Schnabel, ein geöffnetes Maul, kleine Augen, einen langen Hals, einen kleinen Körper und einen flachen Bauch haben, heißen gefiederte Tiere. Sie erscheinen flink und schwach. Ihre Stimme ist rein und durchdringend. Ihre flinke und schwache Erscheinung und ihre reine, durchdringende Stimme passen zu den Klangsteinen. Darum dienen diese Tiere als Säulen für Klangsteingestelle. Schlägt man die aufgehängten Klangsteine an, scheinen die Säulen des Gestells auch mitzuklingen. Tiere mit kleinen,

---

* Siehe Kommentar: Musical Instruments

19 *Die großen Tiere unter dem Himmel bilden fünf Gruppen* – Dieser Abschnitt über die Gestelle zum Aufhängen von Glocken- und Klangsteinspielen enthält eine Klassifikation der Lebewesen. Es ist interessant, dass der nackte Mensch auch als zur Tierwelt zugehörig betrachtet wird. Aber in ausgegrabenen Gestellen sind die Männer, die die Querbalken tragen, bekleidet und tragen Schwerter.

länglichen Köpfen und gleichmäßig rundem Körper heißen geschuppte Tiere. Sie kommen als Figur in den Querhölzern vor.

Bei der Darstellung von wilden Raubtieren muss man die Krallen herausarbeiten, und die Augen müssen vorstehen. Schuppen oder Haare müssen sich aufrichten, dass sie wie jäh in Zorn geraten aussehen. Der jähzornige Ausdruck wird durch das Gewicht der Figur unterstrichen, und sie soll so aussehen, als würde sie kraftvoll brüllen. Wenn die Krallen nicht herausgearbeitet sind, die Augen nicht vorstehen und die Schuppen oder die Haare nicht aufgerichtet sind, dann wirkt die Figur kraftlos und niedergeschlagen. Wirkt sie kraftlos und niedergeschlagen, dann meint man, die Figur könnte die Last nicht tragen. Die Figur wirkt dann matt, und nach ihrem Ausdruck scheint sie nicht zu brüllen.

Der Tischler Ziren 梓人 stellt Trinkgefäße her. Das Volumen von Schöpflöffeln beträgt ein Sheng 升 (0,2 l). Die dreibeinigen Trinkgefäße *jue* 爵 haben ein Volumen von einem Sheng 升 und die Trinkgefäße *gu* 觚 ein Volumen von drei Sheng 升 (0,6 l). Wein wird dem Gast mit dem Trinkgefäß *jue* 爵 überreicht, während man mit dem Trinkgefäß *gu* 觚 einen Toast ausbringt. Man überreicht ein Sheng 升 (0,2 l) Wein und bringt einen Toast mit drei Sheng 升 (0,6 l) Wein aus, was zusammen ein Dou 豆 (0,8 l) ergibt. Man isst ein Dou 豆 Fleisch und trinkt ein Dou 豆 Wein, das ist die Menge von Speisen gewöhnlicher Menschen. Allgemein prüft man die vom Tischler Ziren 梓人 hergestellten Trinkgefäße wie folgt: Wenn man das Gefäß waagerecht zum Mund hebt und in ihm noch Wein zurückbleibt, muss der Aufseher über den Tischler Ziren 梓人 ihn bestrafen.

Der Tischler Ziren 梓人 stellt Schießscheiben her. Breite und Höhe der Schießscheibe bilden ein Quadrat. Das mittlere Feld nimmt ein Drittel der Breite der Schießscheibe ein. Die beiden oben an der Schießscheibe nach rechts und links zeigenden Halterungen sind genauso breit wie die Schießscheibe, so dass die Gesamtbreite das Dreifache der Breite der eigentlichen Schießscheibe beträgt. Die beiden unten angebrachten Halterungen sind halb so breit wie die oben angebrachten. Die Seile an den oberen und unteren Halterungen sind jeweils acht Chi 尺 (157,6 cm) lang. Sie sind an einem Knopf mit einem Durchmesser von einem Cun 寸 (2,0 cm) befestigt. Wenn man eine Schießscheibe aus Leder anbringt, wird das mittlere Feld in der Mitte der Schießscheibe angenäht. Im Frühling wird ein großes Bogenschützenritual abgehalten, um die Fähigkeiten der Lehnsfürsten zu vergleichen. Hierfür wird eine Schießscheibe benutzt, deren Felder mit fünf Farben bemalt sind. Die Lehnsfürsten aus den fernen Reichen versammeln sich bei Hofe. Es wird eine mit Tieren bemalte Schießscheibe aufgespannt. Während der König die Lehnsfürsten bewirtet, wird das Bogenschießenritual abgehalten. Bei diesem Ritual werden Wein, getrocknetes Fleisch und Fleischsoße geopfert.

Der Opferspruch lautet: „Ich nehme mir nur folgsame und verdienstvolle Lehnsfürsten zum Vorbild. Auf keinen Fall dürfen wir uns durch jene ungehorsamen Lehnsfürsten irremachen, die sich nicht im Hof des Königs versammeln und nicht dem Bündnis Folge leisten. Darum werde ich ihnen entgegentreten und sie mit dem Bogen niederschießen. Die folgsamen Lehnsfürsten sollen reichlich zu trinken und zu essen erhalten. Möge großes Glück auf eure Söhne und Enkel übergehen, damit ihr Generation für Generation meine Lehnsfürsten seid!"

25AC. 庐人

庐人为庐器，戈柲六尺有六寸，殳长寻有四尺，车戟常，酋矛常有四尺，夷矛三寻。凡兵，无过三其身，过三其身，弗能用也，而无已，又以害人。故攻国之兵欲短，守国之兵欲长。攻国之人众，行地远，食饮饥，且涉山林之阻，是故兵欲短；守国之人寡，食饮饱，行地不远，且不涉山林之阻，是故兵欲长。凡兵，句兵欲无弹，刺兵欲无蜎，是故句兵椑，刺兵抟。击兵同强，举围欲细，细则校。刺兵同强，举围欲重，重欲傅人，傅人则密，是故侵之。

凡为殳，五分其长，以其一为之被，而围之。参分其围，去一以为晋围。五分其晋围，去一以为首围。凡为酋矛，参分其长，二在前，一在后，而围之。五分其围，去一以为晋围。参分其晋围，去一以为刺围。凡试庐事，置而摇之，以眡其蜎也；炙诸墙，以眡其桡之均也；横而摇之，以眡其劲也。六建既备，车不反覆，谓之国工。

25MC. 庐人

庐人制作兵器之柄。戈的柄长六尺六寸，殳长一丈二尺，用于车战的戟长一丈六尺,[20] 酋矛长二丈，夷矛长二丈四尺。凡是兵器，长度不能超过使用者身高的三倍，超过三倍，就没法用了。非但如此，还会伤及使用者。因此进攻一方的兵器要短一些，防守一方的兵器要长一些。进攻方人多，行军路线长，后勤供应紧张，还有山林之阻，因此兵器要短；防守方人数少，饮食供应充裕，行军距离也短，还没有山林之阻，因此兵器要长。凡是兵器，用来钩杀的不希望它掉转，用于刺杀的希望它坚韧而不柔软挠曲，因此用于钩杀的兵器柄的截面要椭圆形，以便于掌握钩杀方向；用于刺杀的兵器的柄要圆整，以使其不易挠曲。用于撞击的兵器把柄各处要强

---

20 这里"寻"、"常"均为长度单位。所谓"八尺为寻，倍寻为常"，可知一寻等于八尺，一常等于一丈六尺。

度一致，手握之处要稍细一些，稍细易于把持牢固。用于刺杀的兵器把柄各处要强度一致，手握之处要稍粗重些，粗重气势大，容易迫近敌人，迫近就容易击中敌人，重创对方。

凡是制作殳，将其长度均分为五份，以其五分之一处为手握处，截面是圆形。以其圆周长的三分之二为末端鐏的周长，[21] 以鐏周长的五分之四作为殳首与柄连接处的周长。凡是做酋矛的，将其长度分为三份，三分之二在前，三分之一在后，截面是圆形。以其圆周长的五分之四作为末端鐏的周长，以鐏周长的三分之二作为前面矛首的周长。凡要检验长兵器柄的质量，将其后尖插在地上摇动它，观察它是否容易挠曲。将其顶在两墙之间，观察其弯曲时是否均匀。横放在膝上，抓住两头摇动它，以观察其是否强劲。战车上戈、殳、戟、酋矛、夷矛五种兵器具备，再加上旌旗，六种物事齐备，车驰行时它们不歪斜倾倒，能够制作出这种水平的兵器的工匠，可以称之为国家良匠。

25. The Weapon Carpenter

The weapon carpenter manufactures the handles of weapons. The handle of a spear *ge* 戈 is six Chi 尺, six Cun 寸 (130.0 cm) long; the lance *shu* 殳 is one Xun 寻, four Chi 尺 (236.4 cm) long. A chariot halberd is one Chang 常 (315.2 cm) long; the lance *you mao* 酉矛 is one Chang 常, four Chi 尺 (394 cm) long; the lance *yi mao* 夷矛 is three Xun 寻 (473 cm) long. It is impractical if the length of all hand weapons exceeds a men's height threefold. If it exceeds the height threefold, it cannot be handled. More than this, it can also endanger the carrier of the weapon. Therefore, offensive weapons should be short, and defensive weapons long. There are many assailants; they have to march long ways, where there is a shortage of food and drink, and they have to clear hurdles in mountains and forests. Therefore, they need short weapons. There are few defenders; they have sufficient food and drink. They do not have to march long ways, and they do not have to clear hurdles in mountains and forests either. Therefore, they need long weapons. Generally, weapons that kill with a hook should not be easily rotatable, whereas thrust weapons should not bend. Therefore, hook weapons have a handle with a flat cross section, whereas thrust weapons have a handle with a round cross section. The handle of a punch weapon should be equally firm in front and in the back, but the location of the grip should be a little bit slimmer. If the location of the grip is slimmer, one can obtain a better hold of the weapon. The handle of thrust weapons should be equally firm in front and in the back, but the location of the grip must be a little bit thicker.

21 晋，即鐏，套在长兵器柄末端带尖的铜套。兵器可以用鐏插在地上。

If the location of the grip is thicker, one can vehemently attack the enemy, hit him in a precise manner and wound him gravely.

When manufacturing the lance *shu* 殳, the location of the grip is placed at one fifth of the total length of the handle. Its cross section is round, and two thirds of its circumference provide the circumference of the bronze closing cap on the handle. Four fifths of the circumference of the closing cap provide the circumference of the top end of the lance *shu* 殳.

When manufacturing the lance *you mao* 酉矛, the location of the grip lies at one third of the total length of the handle. Its cross section is round, and four fifths of its circumference provide the circumference of the bronze closing cap at the end. Two thirds of the circumference of the bronze closing cap provide the circumference of the location where the knife is fixed onto the handle.

Generally, when testing the quality of long weapons, they are stabbed into the earth and moved back and forth in order to see the degree that they bend. They are put between two walls to see whether they bend back equally. Clamped in this way, the weapon is pressed upwards and downwards to assess its strength. If the five different weapons and the banner are set up on the chariot appropriately and will not fall down while driving, then one can speak of an excellent product.

25. Der Waffentischler

Der Waffentischler stellt die Griffstangen von Waffen her. Die Griffstange eines Speers *ge* 戈 ist sechs Chi 尺 sechs Cun 寸 (130,0 cm) lang, eine Lanze *shu* 殳 ist ein Xun 寻 vier Chi 尺 (236,4 cm) lang. Eine Streitwagenhellebarde ist ein Chang 常 (315,2 cm) lang, die Lanze *you mao* 酉矛 ist ein Chang 常 vier Chi 尺 (394 cm) lang, die Lanze *yi mao* 夷矛 ist drei Xun 寻 (473 cm) lang. Es ist unzweckmäßig, wenn die Länge der Streitwaffen länger als das Dreifache der Körpergröße ist. Übersteigt sie das Dreifache der Körpergröße, ist sie nicht mehr sicher zu handhaben. Nicht nur das, sie kann auch den Träger der Waffe gefährden. Deshalb sollen Angriffswaffen kurz und Verteidigungswaffen lang sein. Die Angreifer sind recht viele, und sie müssen lange Strecken marschieren, es mangelt an Trinken und Essen, außerdem müssen sie Hindernisse in Gebirgen und Wäldern überwinden. Deshalb brauchen sie kurze Waffen. Die Verteidiger sind ziemlich wenige, sie haben ausreichend zu trinken und zu essen. Sie müssen keine langen Strecken marschieren, und sie müssen auch keine Hindernisse in Gebirgen und Wäldern überwinden. Deshalb brauchen sie lange Waffen. Allgemein soll man Waffen, die mit einem Haken töten, nicht leicht drehen können, während Stichwaffen sich nicht durchbiegen sollen. Deshalb haben Hakenwaffen eine Griffstange mit flachem Querschnitt, während Stichwaffen eine Griffstange mit rundem Querschnitt haben. Die Griffstange von Stoßwaffen muss vorn und hinten gleich fest sein, aber die Griffstelle ist etwas schlanker. Wenn die Griffstelle schlanker ist, kann man die Waffe besser halten. Die Griffstange von Stichwaffen muss vorn und hinten gleich fest sein, aber die Griffstelle muss etwas dicker sein. Ist die Griffstelle dicker, kann man ungestüm auf den Feind losgehen, ihn genau treffen und ihn schwer verletzen.

Bei der Herstellung der Lanze *shu* 殳 liegt die Griffstelle bei einem Fünftel der Gesamtlänge der Griffstange. Ihr Querschnitt ist rund, und zwei Drittel ihres Umfangs ergeben den Umfang der bronzenen Abschlusskappe auf der Griffstange. Vier Fünftel des Umfangs der Abschlusskappe ergeben den Umfang des oberen Endes der Lanze *shu* 殳.

Bei der Herstellung der Lanze *you mao* 酉矛 liegt die Griffstelle bei einem Drittel der Gesamtlänge der Griffstange. Ihr Querschnitt ist rund, und vier Fünftel ihres Umfangs ergeben den Umfang der bronzenen Abschlusskappe am Ende. Zwei Drittel des Umfangs der bronzenen Abschlusskappe ergeben den Umfang der Stelle, bei der das Messer an der Griffstange befestigt wird.

Allgemein verfährt man bei der Prüfung der Qualität von langen Waffen, indem man sie in die Erde sticht und sie hin und her bewegt, um den Grad der Verbiegung festzustellen. Man steckt sie zwischen zwei Mauern, um zu sehen, ob sie sich gleichmäßig durchbiegt. So eingespannt, drückt man die Waffe in der Mitte nach oben und unten, um ihre Festigkeit zu beurteilen. Wenn die fünf verschiedenen Waffen und das Banner angemessen auf dem Streitwagen aufgestellt werden und sie beim Fahren nicht umfallen, dann kann man von einem hochqualitativen Produkt sprechen.

26AC. 匠人

匠人建国，水地以县，置槷以县，眂以景。为规，识日出之景与日入之景，昼参诸日中之景，夜考之极星，以正朝夕。

匠人营国，方九里，旁三门。国中九经九纬，经涂九轨。左祖右社，面朝后市，市朝一夫。夏后氏世室，堂修二七，广四修一。五室，三四步，四三尺。九阶。四旁两夹窗。白盛。门堂，三之二；室，三之一。殷人重屋，堂修七寻，堂崇三尺，四阿重屋。周人明堂，度九尺之筵，东西九筵，南北七筵，堂崇一筵。五室，凡室二筵。室中度以几，堂上度以筵，宫中度以寻，野度以步，涂度以轨。庙门容大扃七个，闱门容小扃三个，路门不容乘车之五个，应门二彻三个。内有九室，九嫔居之。外有九室，九卿朝焉。九分其国，以为九分，九卿治之。王宫门阿之制五雉，宫隅之制七雉，城隅之制九雉。经涂九轨，环涂七轨，野涂五轨。门阿之制，以为都城之制。宫隅之制，以为诸侯之城制。环涂以为诸侯经涂，野涂以为都经涂。

匠人为沟洫，耜广五寸，二耜为耦，一耦之伐，广尺深尺，谓之田𤰝；田首倍之，广二尺，深二尺，谓之遂。九夫为井，井间广四尺，深四尺，谓之沟。方十里为成，成间广八尺，深八尺，谓之洫。方百里为同，同间广二寻，深二仞，谓之浍。专达于川，各载其名。

凡天下之地埶，两山之间，必有川焉。大川之上，必有涂焉。凡沟逆地防，谓之不行；水属不理孙，谓之不行。梢沟三十里，而广倍。凡行奠水，磬折以参伍。欲为渊，则句于矩。凡沟必因水埶，防必因地埶。善沟者，水漱之；善防者，水淫之。凡为防，广与崇方，其閷参分去一，大防外閷。凡沟防，必一日先深之以为式，里为式，然后可以傅众力。凡任，索约大汲其版，谓之无任。葺屋参分，瓦屋四分，囷、窌、仓、城，逆墙六分。堂涂十有二分。窦，其崇三尺。墙厚三尺，崇三之。

26MC. 匠人

工匠建造都城，首先要用取水平的方法把地整平，然后用悬垂线的方法垂直树立一个表。观察太阳出来时表的影子。以该表为中心画一个圆，在圆周上标记出太阳刚出与将入地平线时表影与圆周的交点，其连线即为正东西方向。还要白天参考正中午时表的影子的方位，夜晚参考极星所在的方位，用这种方法确定东西南北四向。

工匠营建都城，城是正方形的，每边长九里、开三个城门。城中交通要道九纵九横，每条大道的宽度，可容九轨。左侧是祖庙，右侧是社庙，前面是朝堂，后面是市场。市场和朝堂占地都是一百亩。[22] 夏代的世室，正堂南北进深十四步，宽在进深的基础上加进深的四分之一，共十七步半。堂上有室，共五室，室的进深是三步或四步，室宽再加三尺或四尺。堂的四周一共有九座台阶。五室，每室四个门户，每个门户两侧各一个窗户。用白灰刷墙。门侧之堂进深是整体堂的进深的三分之二，即九步二尺；室的进深是整体堂的进深的三分之一。商代的重屋，堂进深七寻，堂基高三尺，重檐复殿，四面落水。[23] 周代的明堂，以九尺长的竹席为度量单位，东西长九席（八丈一尺），南北长七席（六丈三尺），堂基高一席（九尺）。堂上有五室，每室长宽各一丈八尺。室内以几的长度为度量单位，堂上以竹席为度量单位，宫中以寻为度量单位，野外以步为度

22 “市朝一夫”，夫，古代指从事体力劳动的男子。按古代的井田制，一夫受田百亩，故也称百亩为夫。

23 “四阿重屋”，四阿，郑玄注谓 “四阿若今四注屋”，即屋顶四面落水。四阿重屋，通常释为重檐复殿 顶。

量单位，道路宽度则以轨为度量单位。宗庙的大门宽二丈一尺，[24] 庙中之门宽六尺，寝宫区总门宽一丈六尺五寸，[25] 朝门宽二丈四尺。[26] 寝宫区内有九室，宫内九嫔居住；区外有九室，朝廷九卿在那里处理政事。把一个国家的职事分为九类，九卿分别管理。王宫城门屋脊高度按规制为五丈，[27] 宫墙角楼高七丈，城墙角楼高九丈。城中纵向大道可容九车并行，环城之道可容七车并行，野外之道可容五车并行。王宫城门屋脊的规制，就是都城之制。王宫墙角楼的规制，就是诸侯城墙的规制。王城环城路的宽度，就是诸侯城内道路的宽度，王城野外道路的宽度，就是卿大夫采邑城中的路宽。

匠人制作田间水渠，用的耜宽五寸，二耜为一耦，以一耦为标准，挖成深一尺宽一尺的沟渠，这种沟渠叫做田畎。田地入口处的田畎尺寸要加倍，宽二尺，深二尺，这种规格的沟渠叫遂。按井田制的规格，九夫耕种的田地（每夫耕种一百亩）为一井，井与井之间的水渠宽四尺，深四尺，叫做沟。每边长十里的方田，叫做成，成与成之间的水渠宽八尺，深八尺，叫做洫。边长百里的方田叫同，同与同之间的水渠宽二寻，深二仞，叫做浍。浍直通河流，每条浍都有自己的名字，分别被记录下来。

26. Construction Workers*

Construction workers build cities. On leveled land, cords are fixed on vertical rods and with the stretched cords are used to check that the land is level. They observe the sun's shadow and draw circles. They record the direction of the sun's shadow at noon, and at night, they determine the direction of the Pole Star to lay down the cardinal points.

The construction workers build the king's city. It has an area of nine Li 里 (4.2 km) square, and on each side, three gates will be erected. In the capital, three main streets run in a north-south direction, where each main street has three paths. Three main streets run as well in an east-west direction, where each main street has three paths. The paths in both directions are nine wheel-track widths wide (seven Zhang 丈, two Chi 尺 ≈ 14 m). With regard to the arrangement of the king's palace, the ancestors' temple is situated on the left side, and the earth god's temple on the right side, the palace in front, and the

24 郑玄注云："大扃，牛鼎之扃，长三尺。每扃为一个，七个二丈一尺。……小扃，膷鼎之扃，长二尺。 叁个，六尺。"

25 郑玄注云："路门者，大寝之门。乘车广六尺六寸，五个三丈三尺，言不容者，是两门乃容之。两门乃容之，则此门半之，丈六尺五寸。"

26 郑玄注云："正门谓之应门，谓朝门也。二彻之内八尺，三个二丈四尺。"

27 郑玄注云："雉长三丈，高一丈。度高以高，度广以广。"

* See commentary: Civil Engineering.

market in the back. The market and the palace are hundred paces (100 m) square each. The temple hall of the late rulers of the Xia 夏 dynasty has a depth of fourteen Bu 步 (22 m) and the width of the hall is one quarter larger than its depth. As for the five rooms in the temple, the centrally situated room has a size of four Bu 步 (6.28 m) by four Bu 步, four Chi 尺 (7.09 m), and the four rooms on the corners each measure three Bu 步 (4.71 m) by three Bu 步, three Chi 尺 (5.31 m). A staircase with nine steps leads to the temple. On both sides of the temple hall there are two windows and two corner windows. In the hall, a door is foreseen. The walls are whitewashed with limestone.

Two thirds of the depth of the temple hall is taken up by the central temple. One third of the depth of the central temple is taken up by a temple room.

The houses with a double roof of the people of the Yin 殷 dynasty have a hall in a north-south direction with a depth of seven Xun 寻 (11 m). The base of the hall is three Chi 尺 (59 cm) high. A hipped roof is erected on top.

The audience hall of the people of the Zhou 周 dynasty has a bamboo mat as measurement unit with a length of nine Chi 尺 (177 cm). The width in the east-west direction is nine bamboo mats (15.9 m); the depth in the north-south direction is seven bamboo mats (12.4 m). The hall base is one bamboo mat (1.8 m) high. The hall has five rooms, and the length of each room is two bamboo mats (3.5 m).

Within the rooms, a small table *ji* 几 with a length of three Chi 尺 (59 cm) is used as measurement unit; in the halls, the bamboo mat *yan* 筵 (9 Chi = 177 cm); in the palace, the unit *xun* 寻 (8 Chi 尺 = 157 cm); outside, the step *bu* 步 (8 Chi 尺 = 157 cm); and for the streets, the wheel-track width *gui* 轨 (8 Chi 尺 = 157 cm).

The temple gate can hold seven large cross-beams[28] for carrying the sacrificial vessels *ding* 鼎. The side gate of the temple can hold three small cross-beams[29] for carrying the sacrificial vessels *ding* 鼎. The gate between the outer and inner palace is a little bit narrower than five carriages driving side by side. The gate to the king's palace is as wide as three carriages driving side by side. Nine houses are located in the inner palace, in which the nine concubines live. In the outer palace, nine houses are located in which the nine ministers conduct the affairs of state. The palace takes up one ninth of the king's city. The affairs of the country are divided into nine tasks, which are governed by the nine ministers.

---

28 *Large crossbeams* – The large crossbeam means a carrying pole three Chi 尺 (59.1 cm) long. It is put through the eyes on the vessel. The eyes are vertical handles through which one can put the carrying pole.

29 *Small crossbeams* – Here a small carrying pole two Chi 尺 (39.4 cm) long is meant.

The gate in the palace wall is constructed with a height of five Zhi 雉 (9.8 m); the height of the corner towers of the palace wall is seven Zhi 雉 (13.8 m); and the height of the corner towers of the city wall is nine Zhi 雉 (17.7 m). The width of the streets in a north-south direction is nine wheel-track widths (14 m); the width of the street around the palace wall is seven wheel-track widths (11 m); and the width of the streets in the suburbs is five wheel-track widths (8 m). For the height of the city-wall towers in the vassal cities of the princes and dignitaries, the height of the gate in the palace wall is set (five Zhi 雉 = 9.8 m). For the height of the city wall towers in the cities of the vassal princes, the height of the corner towers in the palace wall is set (seven Zhi 雉 = 13.8 m). For the streets in a north-south direction for the vassal princes, the width of the street around the palace wall is set (seven wheel-track widths = 11 m). For the streets in a north-south direction for the princes and dignitaries, the width of the streets in the suburbs is set (five wheel-track widths = 8 m).

The construction workers dig irrigation channels. Their spade is five Cun 寸 (9.8 cm) wide. Two men dig up the earth with their spades shoulder to shoulder, so that a ditch takes shape. If the excavated earth is one Chi 尺 (20 cm) wide and one Chi 尺 deep, it concerns a rut between the fields. At the beginning of a field, the ditch is twice as wide: two Chi 尺 (39 cm) wide and two Chi 尺 deep; this is called a groove. The fields of nine peasants form a well field[30] with the size of one Li 里 (300 Bu 步 = 471 m) square. The water ditch between the well fields is four Chi 尺 (79 cm) wide and four Chi 尺 deep; it is called a gutter. Between land pieces of ten Li 里 (4.7 km) square, the ditch is eight Chi 尺 (1.57 m) wide and eight Chi 尺 deep; it is called a drain. Between land pieces of one hundred Li 里 (47 km) square, the ditch is two Xun 寻 (3.14 m) wide and two Ren 仞 (3.14 m) deep; it is called a channel. The water of the channel flows into a river. Their names are recorded. The shape of the earth is such that between two mountains there is always a river. Along the large rivers run streets. If one digs a ditch that runs counter to the arteries of the earth, the water cannot flow freely. If the water does not flow in correspondence to the shape of the earth, it will be held up. A water ditch over a stretch of thirty Li 里 (14.1 km) should widen to its double. If one wants to drain off stagnant water, a barrage is made with an obtuse angle, Qingzhe 磬折 (152°), where the ratio of the angle's sides is three to five. If one wants to dig up a water reservoir, rectangular drop steps are built. If one digs up an irrigation ditch, one must follow the

30 *The fields of nine peasants form a well field* – The well-field system, derived from the form of the character *jing* 井 (well), entailed that from nine fields belonging to nine peasants, the field in the middle was cultivated together, and the harvest from this field was given to the ruler as a tax.

current of the water. When building embankments, one must follow the shape of the earth. When digging up a water ditch, one can make use of the current to wash out the earth. In the construction of an embankment, one can use the mud deposited by the water to reinforce it. When building embankments, their upper width equals their height. The slope of the scarp has a ratio of one to one and a half. Wider embankments are reinforced in their base. When digging up ditches and embankments, first gauges of the cross section are made, which are set up at distance intervals of each one Li 里 (471 m). After this, one can employ many workers at the same time. When building walls and embankments, one uses a casing made from wooden boards that are tied together with ropes. If one pulls the ropes excessively, the boards will bend, so that they cannot hold the weight of the earth. It is as if the boards are not tied together.

The height of the roof of thatched houses is a third of their depth. The height of the roof of tiled houses is a quarter of their depth. On round granaries, pits, granaries with a square cross section and city walls, the width of the coping is one sixth of their height. Streets before a stairwell to a hall rise in the middle by one twelfth of their width against their margins. Watercourses in the palace area have a depth of three Chi 尺 (59 cm) in their cross section. The wall around the palace area is three Chi 尺 thick; its height is threefold the wall's thickness.

26. Die Bauleute*

Die Bauleute bauen Städte. Auf dem eingeebneten Boden werden Schnüre an senkrechten Stäben befestigt, um mit den aufgespannten Schnüren die Geradheit zu prüfen. Sie beobachten den Sonnenschatten und zeichnen Kreise. Sie halten den Schatten bei Sonnenaufgang und -untergang fest. Am Tage bestimmen sie die Richtung des Sonnenschattens am Mittag und in der Nacht die Richtung des Polarsterns, um die Himmelsrichtungen festzulegen.

Die Bauleute erbauen die Königsstadt. Sie hat eine Fläche von neun Li 里 (4,2 km) im Quadrat, und an jeder Seite werden drei Tore errichtet. In der Metropole verlaufen drei Hauptstraßen in Nord-Süd-Richtung, wobei jede Hauptstraße drei Wege hat. In Ost-West-Richtung verlaufen auch drei Hauptstraßen, wobei jede Hauptstraße drei Wege hat. Die Wege in beiden Richtungen sind neun Wagenspurbreiten breit (sieben Zhang 丈 zwei Chi 尺 ≈ 14 m). Hinsichtlich der Anordnung des Königspalastes befindet sich auf

* Siehe Kommentar: Civil Engineering.

der linken Seite der Ahnentempel und auf der rechten Seite der Erdgotttempel, vorn der Palast und hinten der Markt. Der Markt und der Palast sind jeweils hundert Schritte (100 m) im Quadrat. In den Palästen der späten Herrscher der Xia 夏-Dynastie hat die Haupthalle eine Tiefe von vierzehn Bu 步 (22 m) und die Breite der Halle ist um ein Viertel größer als die Tiefe. Von den fünf Räumen im Tempel hat der in der Mitte gelegene Raum eine Abmessung von vier Bu 步 (6,28 m) × vier Bu 步 vier Chi 尺 (7,09 m), und die vier Räume auf den Ecken messen jeweils drei Bu 步 (4,71 m) × drei Bu 步 drei Chi 尺 (5,31 m). Zum Tempel führt eine Treppe mit neun Stufen. In der Tempelhalle befinden sich auf jeder Seite zwei Fenster, dazu kommen zwei Eckfenster. In der Halle ist eine Tür vorgesehen. Die Wände werden mit Muschelkalk getüncht.

Die Tiefe der Tempelhalle ist zu zwei Dritteln vom zentralen Tempel eingenommen. Die Tiefe des zentralen Tempels ist zu einem Drittel vom Tempelraum eingenommen.

Die Häuser mit einem doppelten Dach der Menschen der Yin 殷-Dynastie haben eine Tiefe der Halle in Nord-Süd-Richtung von sieben Xun 寻 (11 m). Das Fundament der Halle ist drei Chi 尺 (59 cm) hoch. Oben wird ein Walmdach errichtet.

Die Audienzhalle der Menschen der Zhou 周-Dynastie hat eine Bambusmatte mit neun Chi 尺 (177 cm) Länge als Maßeinheit. Die Breite in Ost-West-Richtung beträgt neun Bambusmatten (15,9 m), die Tiefe in Nord-Süd-Richtung – sieben Bambusmatten (12,4 m). Das Hallenfundament ist eine Bambusmatte hoch (1,8 m). Eine Halle hat fünf Räume, wobei die Länge jedes Raums zwei Bambusmatten (3,5 m) beträgt.

Innerhalb der Räume wird ein Tischchen *ji* 几 (3 Chi 尺 = 59 cm) als Längeneinheit benutzt, in den Hallen die Bambusmatte *yan* 筵 (9 Chi 尺 = 177 cm), im Palast das Maß *xun* 寻 (8 Chi = 157 cm) und im Freien der Schritt *bu* 步 (5 Chi = 98 cm) und bei den Straßen die Wagenspurbreite *gui* 轨 (8 Chi 尺 = 157 cm).

Ein Tempeltor kann sieben große Tragstangen[31] zum Tragen von Opfergefäßen *ding* 鼎 fassen. Das Seitentor eines Tempels kann drei kleine Tragstangen[32] zum Tragen von Opfergefäßen *ding* 鼎 fassen. Das Tor zwischen dem äußeren und dem inneren Palastbereich ist etwas schmaler als fünf nebeneinander fahrende Wagen. Das Tor zum Königspalast ist so breit wie drei nebeneinander fahrende Wagen. Im inneren Palastbereich befinden sich neun Häuser, in denen die neun Konkubinen wohnen. Der äußere Palastbereich

31 *Große Tragstangen* – Die großen Tragstangen waren drei Chi 尺 (59,1 cm) lang. Die Tragstange wurde durch die großen Ösen auf dem Gefäß geschoben.

32 *Kleine Tragstangen* – Die kleinen Tragstangen waren zwei Chi 尺 (39,4 cm) lang.

umfasst neun Häuser, in denen die neun Minister den Staatsgeschäften nachgehen. Die Staatsgeschäfte sind in neun Aufgaben unterteilt, die die neun Minister verwalten.

Das Tor in der Palastmauer ist mit einer Höhe von fünf Zhi 雉 (9,8 m), die Höhe der Ecktürme der Palastmauer mit sieben Zhi 雉 (13,8 m) und die Höhe der Ecktürme der Stadtmauer mit neun Zhi 雉 (17,7 m) festgelegt. Die Breite der Straßen in Nord-Süd-Richtung beträgt neun Wagenspurbreiten (14 m), die Breite der Straße um die Palastmauer herum – sieben Wagenspurbreiten (11 m) und die Breite der Straßen in der Vorstadt – fünf Wagenspurbreiten (8 m). Für die Höhe der Stadtmauertürme in den Lehensstädten der Prinzen und Würdenträger ist die Höhe des Tors in der Palastmauer festgelegt (fünf Zhi 雉 = 9,8 m). Für die Höhe der Stadtmauertürme in den Städten der Lehnsfürsten ist die Höhe der Ecktürme der Palastmauer festgelegt (sieben Zhi 雉 = 13,8 m). Für die Straßen in Nord-Süd-Richtung ist bei den Lehnsfürsten die Breite der Straße um die Palastmauer festgelegt (sieben Wagenspurbreiten = 11 m). Für die Straßen in Nord-Süd-Richtung ist bei den Prinzen und Würdenträgern die Breite der Straßen in der Vorstadt festgelegt (fünf Wangenspurbreiten = 8 m).

Die Bauleute legen Bewässerungskanäle an. Ihr Spaten ist fünf Cun 寸 (9,8 cm) breit. Zwei Männer heben mit den Spaten Schulter an Schulter die Erde aus, so dass ein Graben entsteht. Wenn der Aushub ein Chi 尺 (20 cm) breit und ein Chi 尺 tief ist, handelt es sich um eine Furche zwischen den Feldern. Am Anfang eines Feldes ist der Graben doppelt so groß, zwei Chi 尺 (39 cm) breit und zwei Chi 尺 tief, man nennt ihn Rinne. Die Felder von neun Bauern bilden ein Brunnenfeld[33] mit der Größe von einem Li 里 (300 Bu 步 = 471 m) im Quadrat. Der Wassergraben zwischen den Brunnenfeldern ist vier Chi 尺 (79 cm) breit und vier Chi 尺 tief, man nennt ihn Gosse. Zwischen Landstücken von zehn Li 里 (4,7 km) im Quadrat ist der Graben acht Chi 尺 (1,57 m) breit und acht Chi 尺 tief, er heißt Graben. Zwischen Landstücken von hundert Li 里 (47 km) im Quadrat ist der Graben zwei Xun 寻 (3,14 m) breit und zwei Ren 仞 (3,14 m) tief, man nennt ihn Kanal. Das Wasser des Kanals fließt in einen Fluss. Ihre Namen sind aufgezeichnet. Die Oberflächengestalt der Erde bedingt, dass sich zwischen zwei Gebirgen immer ein Fluss befindet. Entlang der großen Flüsse verlaufen Straßen. Wenn man Gräben anlegt, die den Adern der Erde zuwiderlaufen, kann das Wasser nicht frei strömen. Wenn das Wasser nicht entsprechend der Gestalt der Erde fließt, wird es stocken. Ein Wassergraben sollte sich über eine Strecke von dreißig Li 里 (14 km) auf

33 *Die Felder von neun Bauern bilden ein Brunnenfeld* – Das Brunnenfeldsystem, das sich vom Schriftzeichen *jing* 井 (Brunnen) ableitet, drückt aus, dass von neun Feldern, die neun Bauern gehören, das Feld in der Mitte gemeinsam bestellt und dass die Ernte dieses Felds dem Herrscher als Steuer übergeben wurde.

das Doppelte verbreitern. Wenn man stehendes Wasser ableiten will, legt man eine Staustufe mit einem stumpfen Winkel Qingzhe 磬折 (152°) an, wobei das Verhältnis der beiden Winkelschenkel drei zu fünf ist. Will man ein Wasserreservoir anlegen, dann baut man rechtwinklige Gefällestufen. Wenn man einen Bewässerungsgraben baut, muss man der Strömung des Wassers folgen. Beim Bau von Deichen muss man der Oberflächengestalt der Erde folgen. Bei der Anlage eines Wassergrabens kann man das Auswaschen der Erde durch die Strömung ausnutzen. Bei der Anlage eines Deiches kann man den vom Wasser angeschwemmten Schlamm zu seiner Verstärkung benutzen. Beim Bauen von Deichen sind seine obere Breite und seine Höhe gleich. Die Neigung der Böschung steht im Verhältnis von 1:1,5. Breitere Deiche werden im Fundament verstärkt. Bei der Anlage von Gräben und Deichen muss man zuvor Lehren des Querschnitts anfertigen, die in Abständen von je einem Li 里 (471 m) aufgestellt werden. Danach kann man viele Arbeiter gleichzeitig einsetzen. Beim Bauen von Mauern und Deichen benutzt man eine Verschalung aus Brettern, die mit Seilen zusammengebunden werden. Wenn man die Seile übermäßig anzieht, biegen sich die Bretter durch, so dass sie das Gewicht der Erde nicht halten können. Das ist so, als ob man die Bretter nicht festgebunden hätte.

Die Höhe des Daches von strohgedeckten Häusern beträgt ein Drittel seiner Tiefe. Die Höhe des Daches von ziegelgedeckten Häusern beträgt ein Viertel seiner Tiefe. Bei runden Speichern, Erdmieten, Speichern mit quadratischem Querschnitt und Stadtmauern beträgt die Breite der Krone ein Sechstel der Höhe. Bei Straßen vor dem Treppenaufgang zu einer Halle erhebt sich die Mitte der Straße gegenüber ihren Rändern um ein Zwölftel der Breite von der Straßenmitte zum Straßenrand. Wasserläufe im Palastbereich haben im Querschnitt eine Tiefe von drei Chi 尺 (59 cm). Die Mauer um den Palastbereich ist drei Chi 尺 dick, ihre Höhe beträgt das Dreifache der Dicke der Mauer.

27AC. 车人

车人之事，半矩谓之宣，一宣有半谓之欘，一欘有半谓之柯，一柯有半谓之磬折。

车人为耒，庛长尺有一寸，中直者三尺有三寸，上句者二尺有二寸。自其庛，缘其外，以至于首，以弦其内，六尺有六寸，与步相中也。坚地欲直庛，柔地欲句庛，直庛则利推，句庛则利发。倨句磬折，谓之中地。

车人为车，柯长三尺，博三寸，厚一寸有半。五分其长，以其一为之首。毂长半柯，其围一柯有半。辐长一柯有半，其博三寸，厚三之一。渠三柯者三。行泽者欲短毂，行山者欲长毂。短毂则利，长毂则安。行泽者反輮，行山者仄輮。反輮则易，仄輮则完。六分其轮崇，以其一为之牙围。柏车毂长一柯，其围二柯，其辐一柯，其渠二柯者三。五分其轮崇，以其一为之牙围。大车崇三柯，绠寸。牝服二柯有参分柯之二。羊车二柯有参分柯之一，柏车二柯。凡为辕，三其轮崇，参分其长，二在前，一在后，以凿其钩。彻广六尺，鬲长六尺。

27MC. 车人

车人的工作，涉及角度问题，以直角（矩）为基本单位，直角的一半叫做宣，宣的一倍半叫欘，欘的一倍半叫柯，柯的一倍半叫磬折。

车人制作耒，下端的“庛”长一尺一寸，中间直的部分长三尺三寸，上端勾曲部分长二尺二寸。从下端的“庛”的外缘，沿耒身量起，至上端的勾曲处，共六尺六寸，取直线测量，则与一步的长度相当。坚硬的田地耒的庛要直一些，直庛容易插入土中；柔软的田地耒的庛要勾曲一些，勾曲利于翻地。庛与耒中间的直木部分的夹角如果是一个磬折的话，这种耒适用于不软不硬的田地。

车人制作车，所用斧子的斧柄长三尺，斧头宽三寸，厚一寸半。斧头长度为斧柄长度的五分之一。用于套装车轴的车毂长度为斧柄一半，一尺五寸；车毂周长为斧柄长度的一倍半，四尺五寸。连接车毂与轮圈的辐条长度是斧柄长度的一倍半，四尺五寸，辐条的宽度是三寸，厚一寸。轮圈用三条三倍于斧柄长度的木头揉成。行驶在泥泞的道路上的车，需要短毂，短毂车行驶起来比较灵便；行驶在山间崎岖不平道路上的车，需要长毂，长毂车行驶起来比较稳重。行驶于泥泞道路上的车在揉制轮圈时需要反揉，使木材的心材一面向外，心材致密坚硬，不易沾泥，这样在泥地行驶较为容易。行驶于山地的车在揉制轮圈时需要侧揉，使轮圈的外围同时含有木材的心材和边材，这样可以使轮圈刚柔相济，经久耐用。轮圈的高度是轮高的五分之一。适用于平地载重的大车，车轮高九尺，用于安装辐条的论绠厚一寸。另一种稍小的车，轮高八尺，相较于载重的大车，更适合母牛来驾，叫牝服。比牝服更小一些的是宫中用车，叫羊车，车轮高七尺。柏车比羊车更小，轮高六尺。凡是制作车辕，辕的长度是轮高的三倍，辕的三分之二在车轴的前面，三分之一在车轴的后面，在对应于车轴的位置凿钩衔以连接车轴，轮的间距八尺，[34] 辕端用于压在牛颈上的横木长六尺。

---

34 彻，车的轮距。鬲，车辕前端的横木，用于压在牛颈上，牛通过顶推鬲而驾车前行。彻不可能与鬲同长。诸多治《考工记》的学者均认为此处 “彻广六尺” 是 “彻广八尺” 之误。考察《考工记》他处有关车的轮距的记载，可知当时车的轮距确为八尺。

27. The Carriage Manufacturer*

The work of the carriage manufacturer. A half-right angle (45°) is called Xuan 宣. The angle Xuan 宣 plus its half is called Zhu 欘 (67.5°). The angle Zhu 欘 plus its half is called Ke 柯 (101.25°). The angle Ke 柯 plus its half is called Qingzhe 磬折 (151.875°).

The carriage manufacturer manufactures the grip of the plow. The lower end of the grip is one Chi 尺, one Cun 寸 (21.7 cm) long. The middle, straight grip piece is three Chi 尺, three Cun 寸 (65.0 cm) long. The upper, bent piece is two Chi 尺, two Cun 寸 (43.3 cm) long. In this way, when measuring from the lower end through the bent grip piece up until the upper end, a total length of six Chi 尺, six Cun 寸 (130.0 cm) is found. A direct line from the lower until the upper end is one Bu 步 (98.0 cm) long.[35] If the earth is very hard, the lower end must be straight; if the earth is soft, the lower end should be bent. A straight lower end has the advantage that one easily plows into the earth. A bent lower end has the advantage that it is convenient for digging up clay. If the angle between the lower end and the middle, straight grip piece is one Qingzhe 磬折, the grip is suitable for both soft and hard earth, so that it can be used for every type of earth.

The carriage manufacturer manufactures the carriage. The length of the axe grip Ke 柯 is used as a length unit. An axe grip is three Chi 尺 (59.1 cm) long, three Cun 寸 (5.9 cm) wide and one and a half Cun 寸 (3.0 cm) thick. One fifth of the axe grip's length provides the length of the axe's cutting edge. The length of the carriage hub is half a Ke 柯 (29.6 cm) and the circumference of the carriage hub is one and a half Ke 柯 (88.7 cm). The length of the spokes is one and a half Ke 柯, their width is three Cun 寸 (5,9 cm) and their thickness is one Cun 寸 (2.0 cm). The wheel rim is composed of sections that are each three Ke 柯 (177.3 cm) long. Carriages that drive through marshes have a short hub, whereas carriages that drive through mountains have a long hub. A short hub enables one to drive quickly, whereas a long hub is more stable. For carriages that should drive through marshes, the wheel rim is manufactured in such a way that the outside is heartwood, and the inside sapwood. For carriages that should drive through mountains, the wheel rim is manufactured in such a way that on its outside, both heartwood and sapwood are used. If the heartwood is outside and the sapwood inside, the wheel rim is smooth. If both heartwood

---

* See commentary: Manufacture of Carriages.

35 *A direct line from the lower until the upper end is one Bu 步 (98.0 cm) long* – Here a relation of 1 Bu = 5 Chi seems to be realistic. But in the chapter 26 "Civil engineering" a relation of 1 Bu = 8 Chi is possibly more adequate.

and sapwood are on the outside of the wheel rim, it is quite firm. One sixth of the wheel height provides the circumference of the cross section of the rim. On large carriages for mountains, the carriage hub is one Ke 柯 (59.1 cm) long and the circumference of the wheel hub two Ke 柯 (118.2 cm). The spokes are one Ke long. The wheel rim is composed of three segments that are each two Ke (118.2 cm) long. One fifth of the wheel's height provides the circumference of the rim's cross section. The wheels of large carriages drawn by cows are three Ke 柯 (177.3 cm) large; the inward slope of the spokes is one Cun 寸 (2.0 cm). The carriage body is two and two-thirds Ke 柯 (157.8 cm) long. The carriage body of carriages drawn by goats is two and one-third Ke 柯 (137.7 cm) long. The carriage body of large carriages for the mountains is two Ke 柯 (118.2 cm) long. The pole is manufactured so that it is three times larger than the diameter of the wheel. If the pole is divided into three parts, and two parts are in front and one part behind, a hook for the axle is fixed between the two parts. The distance between the two wheels is eight Chi 尺 (157.6 cm). The yoke of the carriage is six Chi 尺 (118.2 cm) long.

27. Der Wagenbauer*

Die Arbeit des Wagenbauers. Ein halber rechter Winkel (45°) heißt Xuan 宣. Der Winkel Xuan 宣 plus seine Hälfte heißt Zhu 欘 (67,5°). Der Winkel Zhu 欘 plus seine Hälfte heißt Ke 柯 (101,25°). Der Winkel Ke 柯 plus seine Hälfte heißt Qingzhe 磬折 (151,875°).

Der Wagenbauer fertigt den Griff für den Hakenpflug an. Das untere Ende des Griffs ist ein Chi 尺 ein Cun 寸 (21,7 cm) lang. Das mittlere gerade Griffstück ist drei Chi 尺 drei Cun 寸 (65,0 cm) lang. Der obere gekrümmte Teil ist zwei Chi 尺 zwei Cun 寸 (43,3 cm) lang. Vom unteren Ende über das gekrümmte Griffholz bis zum oberen Ende ergibt sich so eine Gesamtlänge von sechs Chi 尺 sechs Cun 寸 (130,0 cm). Die gerade Entfernung vom unteren zum oberen Ende ist ein Bu 步 (98,0 cm) lang.[36] Bei sehr fester Erde muss das untere Ende ganz gerade sein, bei weicher Erde soll das untere Ende gekrümmt sein. Das

* Siehe Kommentar: Manufacture of Carriages.

36 *Die gerade Entfernung vom unteren zum oberen Ende ist ein Bu 步 (98,0 cm) lang* – An dieser Stelle scheint ein Verhältnis 1 Bu = 5 Chi realistisch zu sein. Im Kapitel 26 "Bauleute" kann man dagegen von einem Verhältnis 1 Bu = 8 Chi ausgehen.

gerade untere Ende hat den Vorteil, dass man leicht in die Erde eindringt. Das gekrümmte untere Ende hat den Vorteil, dass es bequem zum Ausheben von Lehm ist. Wenn der Winkel zwischen dem unteren Ende und dem mittleren geraden Griffholz ein Qingzhe 磬折 beträgt, ist der Griff sowohl für weiche als auch harte Erde geeignet, so dass man ihn für jede Erde verwenden kann.

Der Wagenbauer fertigt den Wagen an. Die Länge eines Axtgriffs *Ke* 柯 wird als Längeneinheit benutzt. Ein Axtgriff ist drei Chi 尺 (59,1 cm) lang, drei Cun 寸 (5,9 cm) breit und anderthalb Cun 寸 (3,0 cm) dick. Ein Fünftel der Axtgrifflänge ergibt die Länge der Axtschneide. Die Länge der Wagennabe beträgt ein halbes Ke 柯 (29,6 cm) und der Umfang der Wagennabe anderthalb Ke 柯 (88,7 cm). Die Länge der Speichen beträgt anderthalb Ke, ihre Breite drei Cun 寸 (5,9 cm) und ihre Dicke ein Cun 寸 (2,0 cm). Der Radkranz von großen Wagen wird aus drei je drei Ke 柯 (177,3 cm) langen Teilstücken zusammengesetzt. Wagen, die durch Sümpfe fahren, haben eine kurze Nabe, während Wagen, die durch Gebirge fahren, eine lange Nabe haben. Eine kurze Nabe ermöglicht ein gewandtes Fahren, während eine lange Nabe stabiler ist. Bei Wagen, die durch Sümpfe fahren sollen, fertigt man den Radkranz so, dass sich außen Kernholz und innen Splint befinden. Bei Wagen, die durch Gebirge fahren sollen, sind sowohl Kernholz als auch Splint außen am Radkranz. Wenn sich das Kernholz außen und der Splint innen befindet, ist der Radkranz geschmeidig. Befinden sich sowohl Kernholz als auch Splint außen am Radkranz, ist er recht fest. Ein Sechstel der Höhe des Rades ergibt den Umfang des Querschnitts des Radkranzes. Bei großen Wagen für das Gebirge beträgt die Radnabe ein Ke 柯 (59,1 cm) und der Umfang der Radnabe zwei Ke 柯 (118,2 cm). Die Radspeichen sind ein Ke 柯 lang. Der Radkranz ist aus drei je zwei Ke 柯 (118,2 cm) langen Teilstücken zusammengesetzt. Ein Fünftel der Radhöhe ergibt den Umfang des Querschnitts des Radkranzes. Die Räder großer von Rindern gezogener Wagen sind drei Ke 柯 (177,3 cm) groß, und die Neigung der Speichen nach innen beträgt ein Cun 寸 (2,0 cm). Der Wagenkasten ist zwei zwei Drittel Ke 柯 (157,8 cm) lang. Der Wagenkasten von Ziegenwagen ist zwei ein Drittel Ke 柯 (137,7 cm) lang. Der Wagenkasten bei großen Wagen für das Gebirge ist zwei Ke 柯 (118,2 cm) lang. Die Deichsel wird so gefertigt, dass sie drei Mal größer als der Durchmesser des Rades ist. Wenn man die Deichsel in drei Teile unterteilt, befinden sich zwei Teile vorn und ein Teil hinten, und zwischen den beiden Teilen wird ein Haken für die Achse befestigt. Der Abstand zwischen den beiden Rädern beträgt acht Chi 尺 (157,6 cm). Das Joch des Wagens ist sechs Chi 尺 (118,2 cm) groß.

28AC. 弓人

弓人为弓，取六材必以其时。六材既聚，巧者和之。干也者，以为远也；角也者，以为疾也；筋也者，以为深也；胶也者，以为和也；丝也者，以为固也；漆也者，以为受霜露也。

凡取干之道七:柘为上，檍次之，檿桑次之，橘次之，木瓜次之，荆次之，竹为下。凡相干，欲赤黑而阳声。赤黑则乡心，阳声则远根。凡析干，射远者用埶，射深者用直。居干之道，菑栗不迆，则弓不发。

凡相角，秋𦒻者厚，春𦒻者薄，稚牛之角直而泽，老牛之角紾而昔。疢疾险中，瘠牛之角无泽。角欲青白而丰末，夫角之本，蹙于𦙶而休于气，是故柔。柔故欲其埶也，白也者，埶之征也。夫角之中，恒当弓之畏。畏也者必桡，桡，故欲其坚也。青也者，坚之征也。夫角之末，远于𦙶而不休于气，是故脆。脆故欲其柔也，丰末也者，柔之征也。角长二尺有五寸，三色不失理，谓之牛戴牛。

凡相胶，欲朱色而昔。昔也者，深瑕而泽，紾而抟廉。鹿胶青白，马胶赤白，牛胶火赤，鼠胶黑，鱼胶饵，犀胶黄。凡昵之类不能方。凡相筋，欲小简而长，大结而泽。小简而长，大结而泽，则其为兽必剽，以为弓，则岂异于其兽？筋欲敝之敝，漆欲测，丝欲沈。得此六材之全，然后可以为良。

凡为弓，冬析干而春液角，夏治筋，秋合三材，寒奠体，冰析灂。冬析干则易，春液角则合，夏治筋则不烦，秋合三材则合，寒奠体则张不流，冰析灂则审环，春被弦则一年之事。析干必伦，析角无邪，斲目必荼。斲目不荼，则及其大修也，筋代之受病。夫目也者必强，强者在内而摩其筋，夫筋之所由幨，恒由此作，故角三液而干再液。厚其帤，则木坚；薄其帤，则需，是故厚其液而节其帤，约之。不皆约，疏数必侔。斲挚必中，胶之必均。斲挚不中，胶之不均，则及其大修也，角代之受病。夫怀胶于内而摩其角，夫角之所由挫，恒由此作。凡居角，长者以次需，恒角而短，是谓逆桡，引之则纵，释之则不校。恒角而达，譬如终绁，非弓之利也。今夫茭解中有变焉，故挍；于挺臂中有柎焉，故剽。恒角而达，引如终绁，非弓之利。挢干欲孰于火而无赢，挢角欲孰于火而无燂，引筋欲尽而无伤其力，鬻胶欲孰而水火相得，然则居旱亦不动，居湿亦不动。苟有贱工，必因角干之湿以为之柔，善者在外，动者在内，虽善于外，必动于内，虽善亦弗可以为良矣。

凡为弓，方其峻而高其柎，长其畏而薄其敝，宛之无已应。下柎之弓，末应将兴。为柎而发，必动于𦒻，弓而羽𦒻，末应将发。弓有六材焉，维干强之，张如流水。维体防之，引之中参。维角𤼵之，欲宛而无负弦，引之如环，释之无失体如环。材美，工巧，为之时，谓之参均；角不胜干，干不胜筋，谓之参均；量其力，有三均:均者三，谓之九和。九和之弓，角与干权，筋三侔，胶三锊，丝三邸，漆三斞。上工以有余，下工以不足。

为天子之弓，合九而成规；为诸侯之弓，合七而成规；大夫之弓，合五而成规；士之弓，合三而成规。弓长六尺有六寸，谓之上制，上士服之；弓长六尺有三寸，谓之中制，中士服之；弓长六尺，谓之下制，下士服之。

凡为弓，各因其君之躬、志虑、血气。丰肉而短，宽缓以荼，若是者为之危弓，危弓为之安矢；骨直以立，忿埶以奔，若是者为之安弓，安弓为之危矢。其人安，其弓安，其矢安，则莫能以速中，且不深。其人危，其弓危，其矢危，则莫能以愿中。往体多，来体寡，谓之夹臾之属，利射侯与弋。往体寡，来体多，谓之王弓之属，利射革与质。往体来体若一，谓之唐弓之属，利射深。大和无灂，其次筋角皆有灂而深，其次有灂而疏，其次角无灂。合灂若背手文。角环灂，牛筋蕡灂，麋筋斥蠖灂。和弓毄摩，覆之而角至，谓之句弓。覆之而干至，谓之侯弓。覆之而筋至，谓之深弓。

28MC. 弓人

弓人制作弓，伐取所需的六种材料必须在合适的节令进行。六种材料具备以后，能工巧匠将其有机地组合到一起，制成弓。在六种材料中，干的作用是使弓射得远，角的作用是使弓射出的箭速度快，筋的作用是使箭射得深，胶的作用是把诸种材料粘合在一起，丝的作用是缠绕弓体使弓牢固，漆的作用是抵御风霜雨露。

凡选取弓干材料的规则有七：柘木最好，其次是檍木，再其次是檿桑木，第四等是橘树，第五等是木瓜木，第六等是荆木，第七等是竹子。凡考察弓干用材，希望它颜色黑红，敲击时声音清亮。颜色黑红则靠近树心，声音清亮则离树根远。凡解剖弓干用材，要依据其目的采用不同的解剖方式，想要射得远的，要反着木的长势弯曲；想要射得深，则选取厚直的木材。整治弓干用材时，要注意解剖时不能斜行、不能损伤木材的纹理，这样发弓时弓不会扭伤。

凡是考察制弓所用之角，秋天宰牛取的角厚实，春天宰牛取的角单薄，小牛的角挺直而润泽，老牛的角纹理粗乱。牛有久病则角有内伤，瘦瘠之牛，角无光泽。在选择角时，希望角颜色要青白，角尖要丰满。角的根部接近于脑，受和煦之气，因此柔软。柔软的话希望它自然弯曲，这样可以反其势制弓，而白色是有自然弯曲的征验。角的中部，总是贴在弓臂的弓隈[37]处，弓隈处一定是弯曲的，因为是弯曲的，所以希望它坚韧，而青色是坚韧的征验。角的角尖一端，远离脑部，受不到来自脑部的和煦之气的熏养，因此脆。因为脆，故希望它能够柔韧一些，而角尖丰满，是柔韧

37 弓隈，弓末端与中部之间的弯曲处。

的征验。长度达到了二尺五寸、根部呈白色、中部呈青色、角尖丰满的牛角，其价值可与一头牛相当，是非常难得的牛角。

凡是考察制弓所用的胶材，希望胶颜色朱红，胶丝交错纠结。胶丝交错，形成的沟深，有光泽，交错而成的丝束明显。熬制鹿角制成的胶呈青白色，熬制马皮制成的角赤白色，牛皮熬的胶火红色，鼠皮熬的胶黑色，鱼膘熬的胶像筋腱[38]的颜色，犀牛皮胶黄色。其他动物脂膏之类的胶与上述胶类不能相比。凡考察制弓所用的筋材，希望筋丝匀长、筋束密实而润泽。长出这样筋腱的兽类，必然行动剽疾强悍，用它们的筋腱制成弓，射出的箭岂不同样剽疾强悍？制筋一定要使其筋丝充分舒展，以消除其彼此自相牵制产生的内在筋力。漆要清，[39]丝要尽可能干湿同色。符合以上条件的六种弓材都齐备了，然后才可能制作出良弓来。

凡制作弓，冬天剖解弓干，春天浸渍角材，夏天整治筋腱，秋天则整合胶、漆、丝三材，这样利用寒冬季节确定弓的体形，[40]大寒季节使弓松弛以审视漆纹。冬天治弓干易于定型，春天浸治角容易周到融洽，夏季治筋不会错乱纠结，秋季整合胶、漆、丝三材，可以密致融合；大寒季节确定弓形，后来张弓射箭不会走形变样；冰天雪地时令使弓松弛来审视漆纹，可以更好地判断漆的效果；春季安装弓弦，则一年都可用。剖解弓干必须顺着木材的纹理，析分角材必须端正，消除木材上的结疤必须动作舒缓。如果不舒缓，时间久了，筋会受到损害。[41]因为有结疤的地方比较坚硬，会在里面摩擦筋，时间一长，筋的纠缠断裂在所难免，因此角要浸治三次而干则需要浸治二次。[42]制作弓干时加的垫木太厚，弓就会太硬；加的垫木太薄，弓会软弱，因此需要反复浸治，调节垫木的厚度，用丝和胶密密缠绕。不是弓的所有部位都需要这样缠绕，但其他部位的缠绕都要稀疏均匀。砍削弓干必须到位，敷胶必须均匀。砍削弓干不到位，敷胶不均匀，那么时间一长，角就会受害。胶在里面，与角摩擦，角受到损伤，都是由此引起的。凡处理角，长角要放在弓隈处，[43]如果放在弓隈处的角短了，就会反挠，张弓的时候，角产生不了弹力，射出的箭就不会疾劲。[44]但若角片太长了，一直抵达弓用于张弦的萧头，就像把弓一直系在弓匣里一样，张弓放箭都受碍，对弓不利。张弓时，在弓萧和弓隈相接处会产生形变和弹力，所以射出的箭迅疾；在弓的把臂处两侧贴附骨片，可以增加

38 饵，牲兽的筋腱。

39 测，清的意思。

40 寒奠体，奠，此处读定，义同。

41 修者，久也。弓干上有结疤，其释放弹力就会不均匀，时间长了，其上所敷之筋就容易受到损害。

42 液，浸泡的意思。古人通过浸泡角、干等材料，使其所含胶体溶解，这样再风干这些材料时，其所含水分更容易蒸发，从而有效地降低其含水率。《考工记》称此种方法为"液"。

43 《考工记注疏》云，"长者以次需"，意为"当弓之隈也，长短各称其干，短者居萧。"

44 《考工记注疏》郑玄注："较，疾也。既不用力，放之又不疾。"

把臂的厚实和弹力，使射出的箭剽悍有力。如果角片太长，一直抵达弓的萧头，那就像把弓系在弓匣子里一样，张弓射箭都受碍，对弓不利。用火揉治弓干时要控制火候，使火恰到好处而不要烤得过熟；用火揉治角时也要注意火候，使火恰到好处而不要将其烤烂；舒展筋时要使其充分舒展而无伤其弹力；煮胶要熟，水火要适宜，这样不管是处于潮湿还是干燥的环境，弓都不会变形。如果有那些技艺不高的工匠，一定要趁角、干材料尚未完全干透情况下就用火将其揉治成形，外观很好，内部却不安定，这样虽然外形很好，里面却在变动走形，外观好看也不能成为良弓。

凡制作弓，要使弓两端用于张弦的峻（弓弭处的小立柱）是方形的，弓弣要高，弓的隈角要长，敝要薄，达到这样程度的弓，虽然多次拉弓，弓和弦始终保持良好的响应状态。如果弣比较低，弓对弦的回应将主要由萧来完成。[45] 由萧来应弦，弓隈与弓弣相接处必然会松动，导致引力分布不均匀，角与弓干都会受损。弓有六材，惟有弓干坚强，才会使弓张弓时如流水般顺畅。不用弓的时候，要将弓置于弓檠（弓匣）中，防止变形，这样才会保证拉满弓时弦中点与弓把距离正好三尺。只有把角在弓隈处放正，才能做到张弓的时候弓体不走形扭曲，把弓拉满的时候如环形，放箭后弓体无变形仍如环形。材质精良、做工精巧，制作合时，称之为三均；角与干相得，干与筋相得，称之为三均；测量弓的弹力，也达到了三均。这样的三个三均，称之为九和。达到九和的弓，角与干相得，用筋三侔，用胶三锊，用丝三邸，用漆三斞。技术高超的工匠用这些材料有余，技术低劣的工匠用之则不足。

制作天子用弓，其弧度要满足九张弓连起来能围成一个圆的要求；制作诸侯用弓，要七张弓连起来能围成一个圆；大夫用弓，要五张弓连起来能围成一个圆；士用弓，三张弓连起来能围成一个圆。弓的长度为六尺六寸，称为上制，由高个子佩带；弓的长度为六尺三寸，称为中制，中等个佩带；弓的长度为六尺，称为下制，低个子佩带。

凡制作弓，要因人而异，根据用弓者的体形、心胸、脾性确定弓的特性。对身体肥胖、个子低矮、行动迟缓[46] 的人，要为其制作硬弓，配以柔箭；对身形剽悍、性格刚毅、行为果断的人，要为其制作柔弓，配以疾箭。其人性格和缓，其弓柔、其箭也柔，就不可能以一定速度射中目标，而且即使射中也射不深；人性格急躁，弓硬箭快，则很难按自己的意愿射中目标。弓在松弛状态下外挠的程度高，张弦时内曲程度小，这类弓属于夹弓、臾弓之类，适宜于射箭靶和田猎。反之，弓在松弛状态下外挠程度小，张弦使内曲程度高，这类弓属于王弓之类，适宜于射盾甲和木靶。弓松弛时的外挠程度与张弦时的内曲程度一样，这类弓属于唐弓之类，适宜于射深。最好的弓表面的漆没有皱痕，其次则筋、角处都有漆痕

45 末，弓的末端，指萧。
46 荼，读舒，古舒字。

而且漆痕在其中央，[47] 再其次是多处有稀稀疏疏的漆痕，最下等是仅仅角的中央即隈处无漆痕。[48] 弓的表里漆的皱纹相合，就像人的两手手背相合时肌肤纹理相应一样。角上面的漆痕是环形的，牛筋上的漆痕呈麻点状，麋鹿筋上的漆痕就像尺蠖一样。要使用弓，首先要调试弓，拂去灰尘，观摩其状况。[49] 检查弓的各部分，如果角的质量优良，这叫句弓，箭疾但难以射远；角和弓干均优良，叫侯弓，可以射侯（箭靶）；角、干、筋三者都优良，叫深弓，射深射远皆宜。

28. The Bow Manufacturer*

The bow manufacturer manufactures bows. He gets the six different materials corresponding to the right season. After the six materials are prepared, they are assembled skillfully. The bow serves to shoot the arrow far away. The horn serves to let the arrow fly fast. The string serves to let the arrow penetrate deeply. The glue serves to connect. The silk serves to firmly wrap the bow's body. The lacquer serves to resist humidity. For the bow's body, there are seven sources: the best is cudrania wood, followed by grey oak, wild mulberry tree, mandarin tree, flowering quince and blackthorn; bamboo is the most inferior wood. The material of the bow's body should be red and black, and if one taps the wood, it should emit a clear sound. The red-black colour of the wood should be in the heartwood, and the wood with a clear sound should be far from the root. If one cuts the bow wood, the bow is manufactured, if it should shoot far, against the natural bend of the wood. Bows whose arrows should penetrate deeply are manufactured from thick, straight wood. The essence of processing the bow's wood consists of not hurting the grain of the wood. Then, the bow will not break.

As for the choice of horn, the horn of a cow slaughtered in autumn is thick and firm, whereas the horn of a cow slaughtered in spring is thin. The horn of young cattle is straight and moist, whereas the horn of old cattle is bent and dry. If the cattle were ill for a long time, the horn is decayed on the inside. The horn of emaciated cattle is dull. The colour of the horn should be bluish, and its top should be thick. The root of the horn receives warmth from the head and is therefore soft. Due to its softness, it should be bent. If it is white, this

47 《考工记注疏》曰："'其次筋角皆有灂而深' 者，筋在背，角在隈，皆有灂。但深在其中央，两边 无也。"

48 《考工记注疏》曰："'其次角无灂'，谓隈里无灂，萧头及背有之。"

49 《考工记注疏》曰："和犹调也。毄，拂也。将用弓，必先调之，拂之，摩之。"

* See commentary: Manufacture of Bows.

proves the bend. The middle part of the horn is always glued onto the bend of the bow. This part is always bent. Due to the curvature, it is desirable that this part is firm. If the horn shows a bluish colour, it proves its strength. Because the top of the horn is far away from the head, it does not receive its warmth. Therefore, it is quite brittle. Because it is prone to brittleness, it is desirable that it is flexible. A thick top proves its flexibility. If a horn fulfills the requirements, that is, if it is two Chi 尺, five Cun 寸 (49.3 cm) long, the colour of the root is white, the colour of the middle part is bluish and the top is thick, then the value of the horn is the same as the entire cow.

As for the choice of glue, its colour should be light red and the glue should be dry. Its cracks should be deep, and the glue should shine. The patterns on the surface should overlap with each other. Deer glue is bluish, horse glue is reddish, cattle glue is blazing red and rat glue is black. Fish glue has a colour like muscles and tendons, and rhinoceros glue is yellow. Other kinds of glue cannot be compared with these.

As for the choice of tendons, small pieces in the form of narrow, long strips and larger pieces that are round, and shining moistly are needed. If one gets narrow, long strips and large, round and moist pieces, the animal is certainly quick. If one processes its tendons to make a bow, could it not become the same as a quick animal? During the processing of the tendons one must not stretch and contract them again.

The lacquer should be clear.

The colour of the silk should shine as if it is dipped into water.

If one has prepared these six outstanding materials, a bow of high quality can be manufactured.

In the manufacture of bows, the bow's body is cut in winter. In spring, the horn is soaked; in autumn, one uses the silk, the glue and the lacquer in order to connect the bow's body, horn and tendons to each other. In a cold winter, the bow is laid in a bow case to determine its form. In very severe winter weather, the bow is repeatedly pulled to test its lacquer. If the bow's body is cut in winter, the grain of the wood is even and dense. If the horn is soaked in spring, it gets smooth by itself. If one processes the tendons in summer, they cannot stick together. If one puts the three materials together in autumn, they become firm and dense by themselves. If one determines the form of the bow in winter, the bow will not loose its form when pulled. If one tests the bow's lacquer in severe winter weather, one can observe whether the lacquer traces form rings. In spring, the bowstring is put in place, and after waiting for one year, the manufactured bow can be used.

When cutting the bow's body, one must absolutely follow the grain of the wood. The cut of the horn must not be oblique. One must be cautious when removing knots in the bow's body. If knots in the bow's body are not removed cautiously, the tendons will be damaged after the bow is used for a longer period. Because the knots are quite hard, they rub the tendons and will tear them up. Often, this is the reason for tendons breaking.

The horn must be soaked three times and the bow's body two times. If the wooden reinforcement piece in the middle of the bow is too thick, the bow will be too strong. If the wooden reinforcement piece is too thin, the bow will be too weak. Therefore, one must soak the bow several times and one must adjust the thickness of the wooden reinforcement piece. The area where the bow's body and wooden reinforcement piece connect is wrapped in silk and painted with glue. The remaining areas of the bow must not be wrapped in this way, but the silk must be wrapped evenly and firmly. When cutting the bow's body, one must proceed carefully and steadily. The glue must absolutely be applied evenly. If the bow's body is not cut carefully and the glue is not applied evenly, the horn will be spoiled after extended use of the bow. Bow and glue will rub the horn so that it can break. This is often caused by this effect.

When processing the horn, long pieces of horn are laid on the edge of the bow. If the length of the horn is not sufficient, it can bend. If the bow is drawn, then it is limp and weak. If an arrow is shot, it cannot fly at high speed. If the horn piece is too long, so that it reaches to the bow's end, then it is the same as if the bow lies in the bow case. This will not give the bow an advantage. The joint between the bow's end and the horn on the bow's edge is flexible and elastic. Therefore, when an arrow is shot, it can fly at great speed. In the middle of the bow, the location of the grip is reinforced with horn. In that way, when an arrow is shot, it is very fast. If the horn is too long, so that it reaches to the bow's end, then it is the same as if the bow lies in the bow case. It does not offer any advantage for pulling the string and shooting the arrow. It is correct to bend the bow's body over a flame. The flame must not be too hot. If one wants to bend the bow over a flame appropriately, it must not catch fire. When processing the tendons, they must be stretched as far as possible. One must not stretch and pull them together repeatedly, so that the tensile strength is not lost. When cooking the glue, water and fire must be more or less balanced.

A bow manufactured in this way will not warp, whether it is in a dry or a damp place. Some sloppy workers bend the bow over fire when the horn and the bow's body are not yet dry. From the outside, such a bow looks good, but inside, it is not stable. Although from the outside it looks good, inside it will warp, so that despite its good appearance, no good bow will take shape.

During the manufacture of bows, their ends are made square-shaped, and the grip's location has a rectangular cross section. The pieces of horn on the edges are long; the piece of horn on the inside of the grip is thin. Even if such a bow is drawn many times, it will not slacken. A bow whose cross-section of the grip's location is too flat develops only little force on the grip, and when drawn, the bow can be damaged. If the bow is bent, the joint between the piece of horn and the grip can be damaged. If the joint between the piece of horn and the grip is loose, the bow cannot develop its force. If when drawing the bowstring the bow's ends bend, the horn plates and the bow's body can warp against each other.

Six different materials are used for a bow. Here, the strength of the bow's body is very important. The bow must be drawn as effortlessly as if one floats on running water. If the bow is not used, it is laid in the bow case, so that it will not warp. When drawn, the distance from the string to the bow's grip is approximately three Chi 尺 (59.1 cm). The support provided by the horn plate increases the bow's force, and at drawing the horn plate and the string should not be positioned obliquely to each other. When drawn, the bow has a circular form, and when it relaxes, it maintains its circular form. Excellent material, a careful technique and up-to-date processing are called the three weights.[50] If the horn plate and bow's body and the tendons correspond with each other, it is called the three weights. Furthermore, the drawing force of the bow is tested with a weight of three Jun 钧 (18 kg). The three-times-three weights are called the nine harmonies. On a bow of the nine harmonies, the horn plate and the bow's body are adjusted to each other; three Mou 侔 tendons, three Lüe 锊 glue, three Di 邸 silk and three Yu 斞 lacquer are used. For more demanding processing, a little bit more is used, for less demanding processing, a little bit less is used. If a bow is manufactured for the son of heaven, the bow's length amounts to one ninth of the circumference; on a bow for a vassal prince, the bow's length amounts to one seventh of the circumference; on a bow for a dignitary, one fifth of the circumference; and on a bow for an official, one third of the circumference. Bows with a length of six Chi 尺, six Cun 寸 (130.0 cm) are called top category. They are carried by vassals of the highest rank. Bows with a length of six Chi 尺, three Cun 寸 (124.1 cm) are called medium category. They are carried by employees of a medium rank. Bows with six Chi 尺 (118.2 cm) are called lower category. They are carried by employees of a lower rank.

Generally, when manufacturing a bow, the height, willpower, character and temperament of the carrier are considered. Stocky persons who think indulgently and act slowly need a strong, fast bow, and with this strong, fast bow, a soft arrow fits. Persons who act boldly, determined, irritated and swift, need a soft bow, and with this soft bow, a fast arrow fits. If a person is indulgent and

50 *Excellent material, a careful technique and up-to-date processing are called the three weights* – The three weights are here understood as quality criteria of a bow.

slow and he uses a soft bow and a slow arrow, the arrow does not fly fast, and he does not easily hit his aim. Even if he has hit the aim, the arrow does not have the strength to penetrate. If a person is bold and determined and has a hot-tempered character and he uses a strong, fast bow and a swift arrow, naturally he cannot hit the aim safely. If a bow that is not drawn is bent strongly, then the inner curvature when drawing the string is small. Such a bow belongs to the category of weak bows; it is ideal for target shooting and hunting. If on the other hand a bow that is not drawn is only bent a little, and the inner curvature when drawing the string is large, it belongs to the category of king's bows; it is ideal for shooting at shields and armors and wooden targets. If the outer bend of a bow that is not drawn equals the inner curvature when drawing the string, such a bow belongs to the category of medium strong bows, the arrows of which penetrate deeply. The lacquer on the surface of the best bows does not show ridges. These are followed by bows on whose strings and horn parts lacquer ridges appear that are the deepest between both parts. These are followed by bows that show lacquer ridges in many places. The bows of the lowest category show lacquer ridges, except in the middle of the horn and in the bend. The lacquer ridges on the surface of the bow resemble the skin lines on the back of the hand and the palm. The lacquer ridges on the horn part are circular; the lacquer ridges on a cow's tendons look like sesame grains; the lacquer ridges on deer tendons equal a mulberry silkworm moth. If one uses a bow, first one has to check it, wipe off the dust and examine its shape. If the quality of the horn parts is excellent when testing a bow, it is called an average bow if the arrow is fast, but does not fly far. If the horn parts and the bow's body are excellent, such a bow is called a target bow, because one can shoot at a target with it. If the three parts, horn, string and bow's body, are excellent, it is called a deep bow, because it is ideal for shooting far and intruding deeply.

28. Der Bogenmacher*

Der Bogenmacher fertigt Bögen an. Er gewinnt die sechs verschiedenen Materialien entsprechend der richtigen Jahreszeit. Sind die sechs Materialien vorbereitet, werden sie mit Geschick zusammengefügt. Der Bogen dient dazu, den Pfeil weit zu schießen. Das Horn dient dazu, den Pfeil schnell fliegen zu lassen. Die Sehnenbänder dienen dazu, den Pfeil tief eindringen zu lassen. Der Leim dient zum Verbinden. Die Seide dient dazu, den Bogenkörper

* Siehe Kommentar: Manufacture of Bows.

fest zu umwickeln. Der Lack dient dazu, der Feuchtigkeit zu widerstehen. Für den Bogenkörper gibt es sieben Quellen: Das beste ist Kudranienholz, danach folgt die Graueiche, danach der wilde Maulbeerbaum, danach das Holz des Mandarinenbaums, danach der Melonenbaum, danach folgt das Schlehdornholz, und Bambus ist das minderwertigste Material. Das Material für den Bogenkörper soll rot und schwarz sein, und wenn man gegen das Holz schlägt, soll es einen hellen Klang geben. Die rotschwarze Farbe des Holzes soll im Kernholz sein, und der helle Klang soll weit von der Wurzel sein. Wenn man das Bogenholz schneidet, fertigt man den Bogen, wenn er weit schießen soll, entgegen der natürlichen Krümmung des Holzes. Bögen, deren Pfeile tief eindringen sollen, fertigt man aus einem dicken, geraden Holz. Das Wesen bei der Bearbeitung des Bogenholzes besteht darin, beim Schneiden die Maserung nicht zu verletzen. Dann wird der Bogen nicht brechen.

Bei der Auswahl des Horns ist bei einem im Herbst geschlachteten Rind das Horn dick und fest, während das Horn bei einem im Frühling geschlachteten Rind dünn ist. Das Horn eines jungen Rindes ist gerade und feucht, während das Horn eines alten Rindes gekrümmt und trocken ist. Wenn das Rind lange krank war, ist das Horn innen schadhaft. Das Horn eines abgemagerten Rindes ist glanzlos. Die Farbe des Horns soll bläulich und seine Spitze dick sein. Die Wurzel des Horns erhält Wärme vom Kopf und ist deshalb weich. Wegen ihrer Weichheit soll sie gebogen sein. Wenn sie weiß ist, gilt dies als Nachweis der Krümmung. Den mittleren Teil des Horns klebt man immer an die Krümmung des Bogens. Dieser Teil wird immer gebogen. Wegen der Krümmung möchte man, dass dieser Teil fest ist. Wenn das Horn eine bläuliche Farbe zeigt, ist dies ein Nachweis für die Festigkeit. Da die Spitze des Horns weit vom Kopf entfernt ist, erhält sie nicht seine Wärme. Deshalb ist sie recht spröde. Weil sie zur Sprödigkeit neigt, möchte man, dass sie biegsam ist. Eine dicke Spitze ist ein Nachweis für ihre Biegsamkeit. Wenn ein Horn die Anforderungen erfüllt, dass es zwei Chi 尺 fünf Cun 寸 (49,3 cm) lang, die Farbe der Wurzel weiß, die Farbe des mittleren Bereichs bläulich und die Spitze dick ist, dann ist der Wert des Horns so viel wert wie das ganze Rind.

Bei der Auswahl des Leims soll seine Farbe hellrot und der Leim trocken sein. Seine Risse sollen tief sein, und der Leim soll glänzen. Die Muster auf der Oberfläche durchdringen sich. Hirschleim ist bläulich, Pferdeleim rötlich, Rinderleim feuerrot und Rattenleim schwarz. Fischleim hat eine Farbe wie Muskeln und Sehnen, und Rhinozerosleim ist gelb. Andere Leimarten können sich mit diesen nicht vergleichen.

Bei der Auswahl der Sehnen braucht man schmale, lange Streifen und größere Stücke, die rund und feucht scheinend sind. Wenn man schmale, lange Streifen und große, runde, feuchte Stücke erhält, dann ist solch ein Tier bestimmt flink. Wenn man seine Sehnen zur Herstellung eines Bogens benutzt, wie könnte er nicht ebenso werden wie ein flinkes Tier? Bei der Behandlung der Sehnen darf man sie nicht wieder strecken und zusammenziehen.

Der Lack soll klar sein.

Die Farbe der Seide soll glänzen, als ob sie in Wasser getaucht ist.

Wenn man diese sechs hervorragenden Materialien vorbereitet hat, kann man einen hochqualitativen Bogen herstellen.

Bei der Herstellung von Bögen schneidet man den Bogenkörper im Winter zu. Im Frühling weicht man das Horn ein, im Sommer bearbeitet man die Sehnen, im Herbst benutzt man die Seide, den Leim und den Lack, um den Bogenkörper, das Horn und die Sehnen miteinander zu verbinden. Im kalten Winter legt man den Bogen in einen Bogenkasten, um seine Form zu bestimmen. Bei sehr strengem Winterwetter spannt man den Bogen wiederholt, um den Lack auf dem Bogen zu prüfen. Wenn man den Bogenkörper im Winter schneidet, ist die Maserung des Holzes ebenmäßig und dicht. Wenn man das Horn im Frühling einweicht, wird es von selbst geschmeidig. Wenn man die Sehnen im Sommer bearbeitet, können sie nicht zusammenkleben. Wenn man im Herbst die drei Materialien zusammenfügt, werden sie von selbst fest und dicht. Wenn man im Winter die Gestalt des Bogens bestimmt, wird er beim Spannen des Bogens nicht die Form verlieren. Wenn man bei strengem Winterwetter den Bogenlack prüft, kann man feststellen, ob die Lackspuren Ringe bilden. Im Frühling zieht man die Bogensehne auf, und nachdem man ein Jahr gewartet hat, kann man den so hergestellten Bogen benutzen.

Beim Schneiden des Bogenkörpers muss man unbedingt der Maserung des Holzes folgen. Der Schnitt des Horns darf nicht schräg erfolgen. Beim Entfernen der Äste im Bogenkörper muss man behutsam vorgehen. Wenn man beim Entfernen der Äste im Bogenkörper nicht behutsam vorgeht, werden die Sehnen nach längerem Gebrauch des Bogens durch sie beschädigt. Da die Äste recht hart sind, reiben diese harten Äste an den Sehnen und werden sie zerreißen. Oft ist das die Ursache, wenn die Sehnen versagen.

Das Horn muss man dreimal und den Bogenkörper muss man zweimal einweichen. Wenn das Verstärkungsholz in der Mitte des Bogens zu dick ist, wird der Bogen zu fest. Ist das Verstärkungsholz zu dünn, wird der Bogen zu schwach. Darum muss man den Bogen mehrfach einweichen, und man muss die Dicke des Verstärkungsholzes abstimmen. Der Bereich der Verbindung zwischen Bogenkörper und Verstärkungsholz wird mit Seide umwickelt und mit Leim bestrichen. Die übrigen Bereiche des Bogens muss man nicht derart umwickeln, aber die Seide muss gleichmäßig fest gewickelt werden. Beim Zuschneiden des Bogenkörpers muss man sorgfältig und gleichmäßig vorgehen. Der Leim muss unbedingt gleichmäßig aufgetragen werden. Wenn der

Bogenkörper nicht sorgfältig zugeschnitten und der Leim nicht gleichmäßig aufgetragen wird, wird das Horn bei längerem Gebrauch des Bogens versagen. Bogen und Leim reiben am Horn, so dass dieses brechen kann. Das wird oft hierdurch verursacht.

Bei der Verarbeitung des Horns legt man lange Hornstücke auf den Rand des Bogens. Wenn die Länge des Horns nicht ausreicht, kann es sich verbiegen. Spannt man den Bogen, dann ist er schlaff und kraftlos. Schießt man einen Pfeil ab, kann er nicht schnell fliegen. Ist das Hornstück zu lang, so dass es bis zum Bogenende reicht, dann ist das so, als ob der Bogen im Bogenkasten liegt. Der Bogen erlangt dadurch keinen Vorteil. Die Verbindungsstelle zwischen dem Bogenende und dem Horn auf dem Bogenrand ist biegsam und elastisch. Darum kann der abgeschossene Pfeil schnell fliegen. In der Mitte des Bogens ist die Griffstelle mit Horn verstärkt. Dadurch ist der abgeschossene Pfeil so schnell. Wenn das Horn zu lang ist und bis zum Bogenende reicht, dann ist das so, als ob der Bogen im Bogenkasten liegt. Es bietet keinen Vorteil für das Spannen der Sehne und das Abschießen des Pfeils. Es ist richtig, den Bogenkörper über einer Flamme zu biegen. Die Flamme darf nicht zu heiß sein. Wenn man den Bogen über einer Flamme richtig biegen will, darf man ihn nicht anbrennen. Bei der Verarbeitung der Sehnen muss man sie so weit wie möglich dehnen. Man darf sie nicht wiederholt ziehen und entspannen, damit ihre Zugkraft nicht verloren geht. Beim Kochen des Leims müssen sich Feuer und Wasser die Waage halten.

Ein auf diese Weise hergestellter Bogen, ganz gleich ob er sich an einem trockenen oder feuchten Ort befindet, wird sich nicht verziehen. Manche schludrige Arbeiter biegen den Bogen über dem Feuer, wenn das Horn und der Bogenkörper noch nicht trocken sind. Äußerlich sieht solch ein Bogen gut aus, aber im Innern ist er nicht stabil. Obwohl er äußerlich gut aussieht, wird er sich aber im Innern verziehen, so dass aus ihm trotz seines guten Aussehens kein guter Bogen entstehen kann.

Bei der Herstellung von Bögen werden seine Enden quadratisch gestaltet, und die Griffstelle hat einen rechteckigen Querschnitt. Die Hornstücke auf den Rändern sind lang, das Hornstück an der Innenseite des Griffs ist dünn. Ein solcher Bogen wird, auch wenn man ihn viele Mal spannt, nicht erschlaffen. Ein Bogen, bei dem die Griffstelle zu tief liegt, entwickelt am Griff nur eine geringe Kraft, und er kann beim Spannen des Bogens lädiert werden. Wenn der Griff gebogen ist, wird die Verbindungsstelle zwischen Hornstück und Griff beim Spannen des Bogens zerstört. Wenn die Verbindung zwischen Hornstück und Griff zerstört wird, kann sich die Kraft des Bogens nicht entfalten. Wenn sich die Bogenenden beim Spannen der Bogensehne biegen, können sich dann die Hornplatten und der Bogenkörper gegeneinander verziehen.

Für einen Bogen werden sechs verschiedene Materialien eingesetzt. Dabei ist die Stärke des Bogenkörpers sehr wichtig. Der Bogen muss sich so leicht spannen lassen, als ob man sich von fließendem Wasser treiben lässt. Wird der Bogen nicht benutzt, legt man ihn in den Bogenkasten, damit er sich nicht verzieht. Beim Spannen beträgt der Abstand von der Sehne bis zum Bogengriff etwa drei Chi 尺 (59,1 cm). Die Unterstützung durch die Hornplatte vermehrt die Bogenkraft, wobei die Hornplatte und die Sehne beim Spannen nicht schräg zueinander liegen sollen. Der Bogen hat beim Spannen eine Kreisform, und auch wenn er sich entspannt, behält er die Kreisform. Vorzügliches Material, eine sorgfältige Technik und die zeitgerechte Verarbeitung heißen die drei Gewichte.[51] Die Harmonie zwischen Hornplatte und Bogenkörper und zwischen Bogenkörper und Sehnen heißen auch die drei Gewichte. Bei der Prüfung der Bogenkraft hängt man ein Gewicht von drei Jun 钧 (18 kg) an die Sehne, das sind wieder drei Gewichte. Die drei mal drei Gewichte heißen die neun Harmonien. Bei einem Bogen der neun Harmonien sind Hornplatte und Bogenkörper aufeinander abgestimmt, verwendet man drei Mou 侔 Sehnen, drei Lüe 锊 Leim, drei Di 邸 Seide und drei Yu 斞 Lack. Bei anspruchsvoller Verarbeitung verwendet man etwas mehr, bei weniger anspruchsvoller Verarbeitung etwas weniger. Wenn man einen Bogen für den Himmelssohn herstellt, beträgt die Bogenlänge ein Neuntel eines Kreisumfangs, bei einem Bogen für einen Lehnsfürsten beträgt die Bogenlänge ein Siebtel eines Kreisumfangs, bei einem Bogen für einen Würdenträger – ein Fünftel eines Kreisumfangs und bei einem Bogen für einen Beamten – ein Drittel eines Kreisumfangs. Bögen mit einer Länge von sechs Chi 尺 sechs Cun 寸 (130,0 cm) heißen obere Kategorie. Sie werden von Lehnsleuten des höchsten Rangs getragen. Bögen mit einer Länge von sechs Chi 尺 drei Cun 寸 (124,1 cm) heißen mittlere Kategorie. Sie werden von Bediensteten des mittleren Rangs getragen. Bögen mit sechs Chi 尺 (118,2 cm) Länge heißen untere Kategorie. Sie werden von Bediensteten des unteren Rangs getragen.

Allgemein berücksichtigt man bei der Herstellung eines Bogens die Körpergröße, die Willensstärke, den Charakter und das Temperament des Trägers. Untersetzte Personen, die nachsichtig denken und langsam handeln, benötigen einen starken, schnellen Bogen, und zu diesem starken, schnellen Bogen passt ein weicher Pfeil. Personen, die kühn, entschlossen und gereizt und rasch handeln, benötigen einen weichen Bogen, und zu diesem weichen Bogen passt ein schneller Pfeil. Wenn eine Person nachsichtig und langsam ist und sie

---

51 *Vorzügliches Material, eine sorgfältige Technik und die zeitgerechte Verarbeitung heißen die drei Gewichte* – Die drei Gewichte werden hier als Qualitätskriterien für einen Bogen verstanden.

einen weichen Bogen und einen langsamen Pfeil benutzt, dann fliegt der Pfeil nicht schnell, und er trifft nicht leicht sein Ziel. Selbst wenn er das Ziel traf, hat er aber keine Kraft einzudringen. Wenn eine Person kühn und entschlossen ist und einen jähzornigen Charakter hat und sie einen starken, schnellen Bogen und einen flinken Pfeil benutzt, kann sie natürlich das Ziel nicht sicher treffen. Wenn ein Bogen ungespannt stark gebogen ist, dann ist die innere Krümmung beim Spannen der Sehne gering. Ein solcher Bogen gehört zur Art der schwachen Bögen, er eignet sich zum Scheibenschießen und zur Jagd. Wenn demgegenüber die Biegung eines ungespannten Bogens klein und die innere Krümmung beim Spannen der Sehne groß ist, gehört ein solcher Bogen zur Art der Königsbögen; er eignet sich zum Schießen auf Schild und Panzer und hölzerne Schießscheiben. Wenn die äußere Biegung eines ungespannten Bogens gleich der inneren Krümmung beim Spannen der Sehne ist, gehört ein solcher Bogen zur Art der mittelstarken Bögen, deren Pfeile tief eindringen. Der Lack auf der Oberfläche der besten Bögen zeigt keine Riefen. Danach folgen Bögen, bei denen sich an den Sehnen und den Hornteilen Lackriefen zeigen, die zwischen beiden Teilen am tiefsten sind. Danach kommen Bögen, die an vielen Stellen vereinzelte Lackriefen zeigen. Die Bögen der untersten Kategorie zeigen nur in der Mitte des Horns und in der Biegung keine Lackriefen. Die Lackriefen auf der Oberfläche des Bogens gleichen den Hautriefen auf dem Handrücken und dem Handteller. Die Lackriefen auf dem Hornteil sind kreisförmig, die Lackriefen auf Rindersehnen sehen wie Sesamkörnchen aus, die Lackriefen auf Hirschsehnen gleichen einem Maulbeerspanner. Wenn man einen Bogen benutzt, muss man ihn zuerst untersuchen, den Staub abwischen und seine Gestalt prüfen. Wenn die Qualität der Hornteile bei der Kontrolle eines Bogens vorzüglich ist, nennt man ihn einen mittelmäßigen Bogen, wenn der Pfeil zwar schnell ist, aber nicht weit fliegt. Wenn die Hornteile und der Bogenkörper vorzüglich sind, nennt man einen solchen Bogen einen Schießscheibenbogen, denn man kann mit ihm auf eine Schießscheibe schießen. Wenn die drei Teile Horn, Sehne und Bogenkörper vorzüglich sind, nennt man ihn einen tiefen Bogen, weil er geeignet ist, weit zu schießen und tief einzudringen.

PART 2

Analysis – Commentaries

1 Manufacture of Carriages

As the manufacture of a car nowadays represents a high level of technology, the manufacture of chariots in the Chunqiu 春秋 and Zhanguo 战国 periods was a confluence of the arts of different craftsmen.

For carriages, the "Kao Gong Ji" lays down all essential sizes in a modular way. First, six differential sizes related to the charioteer are listed, which are important for the use of the carriage as a war chariot:

TABLE 1 Differential sizes of a carriage

No.	Component (or person)	Height	Differential size
1	Rear railing	4 Chi 尺	4 Chi (above ground)
2	Lance grip	4 Chi 尺	4 Chi (above rear transverse rod)
3	Charioteer	8 Chi 尺	4 Chi (charioteer minus lance)
4	Weapon *shu* 殳	12 Chi 尺	4 Chi (weapon *shu* 殳 minus charioteer)
5	Halberd	16 Chi 尺	4 Chi (halberd minus weapon *shu* 殳)
6	Spear	20 Chi 尺	4 Chi (spear minus halberd)

The structure of a carriage from the time between the Chunqiu 春秋 and Zhanguo 战国 periods, and the designations of the main components, is shown in Fig. 13.

The main components of a carriage are the wheels, which must bear the load of the carriage and rotate smoothly. Fig. 14 depicts the structure of a wheel.

The figure makes clear that the rim was assembled using several, most often three or four, bent parts. The spokes are somewhat inclined to the inner side. Such a construction enabled a better absorption of the forces acting on the outer side (for instance if one rounded a curve).

The sizes of the wheel are provided in Fig. 15. According to the description in "Kao Gong Ji", only the outer part of the wheel rim, which amounted to one third of the rim's thickness due to wear and dust, was not lacquered. The remaining lacquered region was then taken as the basis for the dimensions of the hub.

 | DOI:10.1163/9789004416949_005

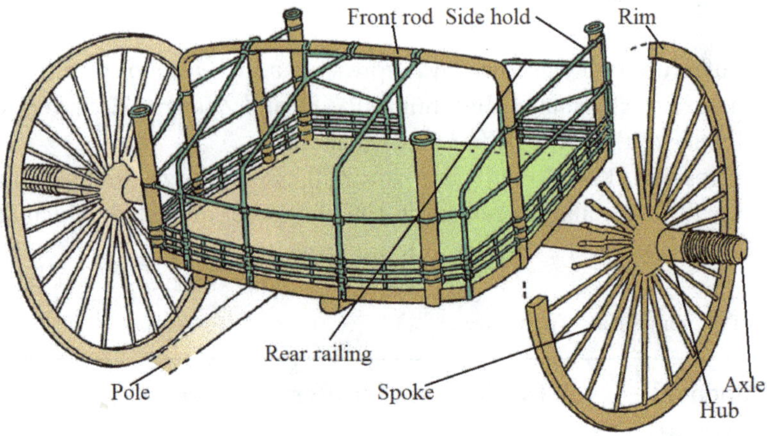

FIGURE 13 Reconstruction of a chariot from tomb No. 8 belonging to Zhao Qing 赵卿 from the state Jin 晋 in Jinsheng 金胜 (now near Taiyuan 太原)
Source: Wenren Jun 闻人军, *Kao Gong Ji yizhu* 考工记译注 (Translated and commented edition of Kao Gong Ji), Shanghai guji chubanshe 上海古籍出版社, 2008, pp. 30–31.

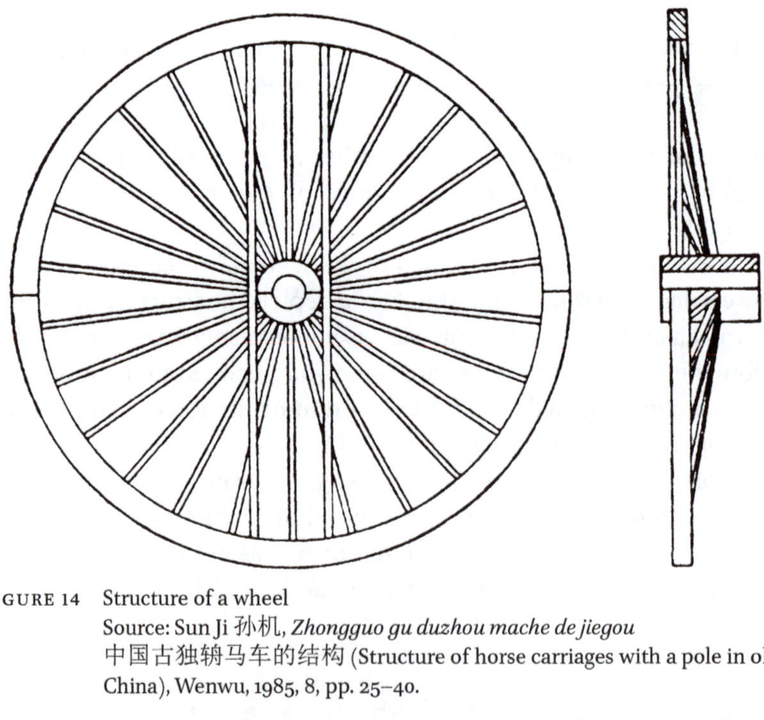

FIGURE 14 Structure of a wheel
Source: Sun Ji 孙机, *Zhongguo gu duzhou mache de jiegou* 中国古独辀马车的结构 (Structure of horse carriages with a pole in old China), Wenwu, 1985, 8, pp. 25–40.

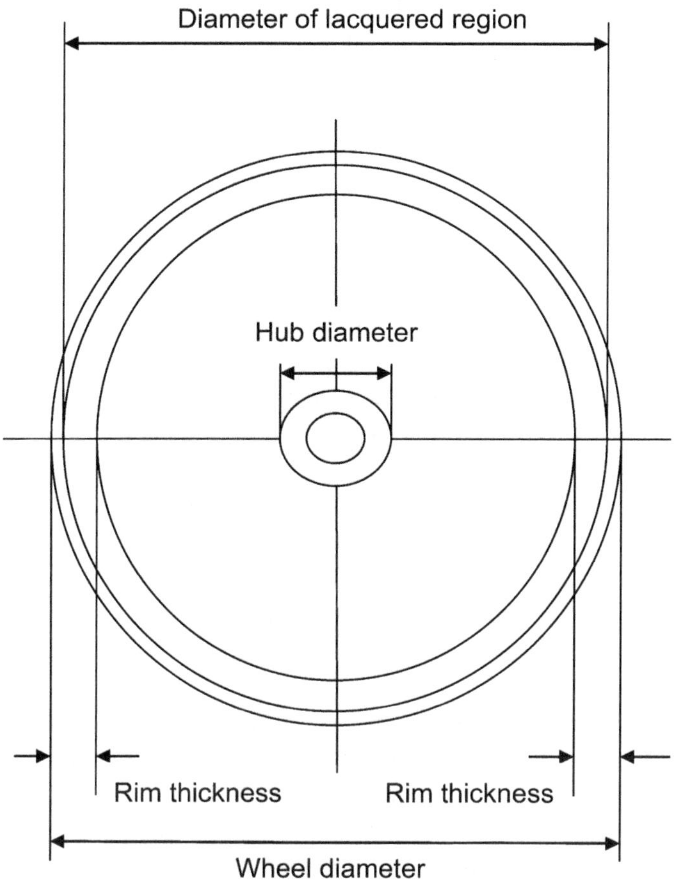

FIGURE 15 Sizes of the wheel

TABLE 2 Sizes of the wheel

Parameter	Size
Diameter of chariots' wheels	6 Chi 尺, 6 Cun 寸
Thickness of wheel rim	(1/6)/2 diameter of the wheel
Diameter of the wheel's lacquered region	= wheel diameter – (2 wheel rim thickness)/3
Length of the hub	= 1/2 diameter of the wheel's lacquered region

Source: Guan Zengjian 关增建, *On the Problem of Sizes of the Hub in Kao Gong Ji*, personal communication, 2011.

The hub is a component of the wheel that is under a lot of stress. The following sizes were laid down for it, provided once more in a modular way.

TABLE 3 Sizes of the hub

Parameter	Size
Circumference of the hub	= length of the hub
Diameter of hub hole	= 1/3 outer diameter of the hub
Depth of pinhole in the hub	= width of the spokes
Circumference of hub hole on the inner side	= 4/5 hub length
Circumference of hub hole on the outer side	= 2/5 hub length

The carriages had a fixed axle. The wheel, with the hub in the centre, would be placed on the axle. Excavations of chariots from the Chunqiu 春秋 and Zhanguo 战国 periods confirmed that the hub hole tapered somewhat towards the outer side (but this was not the case for all excavated chariots). The idea behind this measure was obviously that the end of the axle wore down as the chariot was used, and tended to take on a conical shape.

"Kao Gong Ji's" description of the circumference of the hub hole on the inner side as amounting to four fifths of the hub length, and on the outer side to two fifths of the hub length is not realistic, but hub holes with a light conical shape were discovered on several chariots. The statement that the diameter of the hub hole amounts to one third of the hub's outer diameter contradicts the diameters of the hub's conical hole that were provided. It is possible that the text is erroneous here.

In order for the wheel not to come off the axle, a wooden or metal cap was put on the end of the axle and fixed with a nail. A more elegant solution was an axle pin, which was shaped in the form of a man (see Figs. 16 and 17).

The axle caps were mounted on the end of the axle and had a groove-shaped opening in which the axle end's pin was placed, so that the axle cap was fixed onto the axle.

The "Kao Gong Ji" describes a wheel as having 30 spokes. This number is also mentioned in the "Dao De Jing" 道德经, where it is written: "Thirty spokes surround the hub".[1] But only by very few discovered carriages could confirm this number. More often, the wheels had 26 spokes.[2]

1 Laudse, *Daudedsching* (Laozi, Dao De Jing), Reclam Verlag Leipzig, 1970, p. 67.

2 Sun Ji, *Structure of Horse Chariots with a Pole in Ancient China*, Wen Wu, 1985, 6, pp. 25–40.

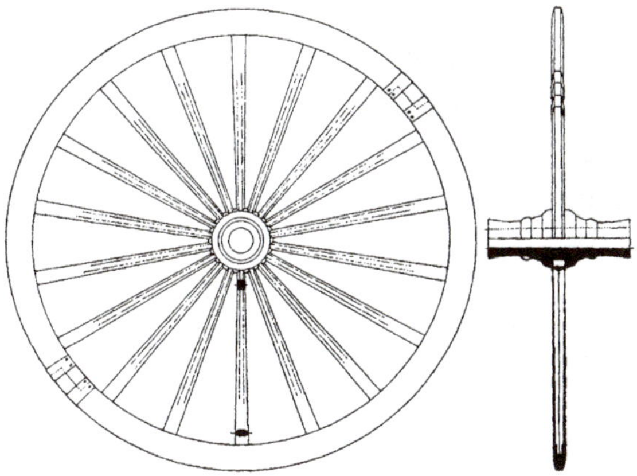

FIGURE 16A Shows the hub of the wheel in more detail

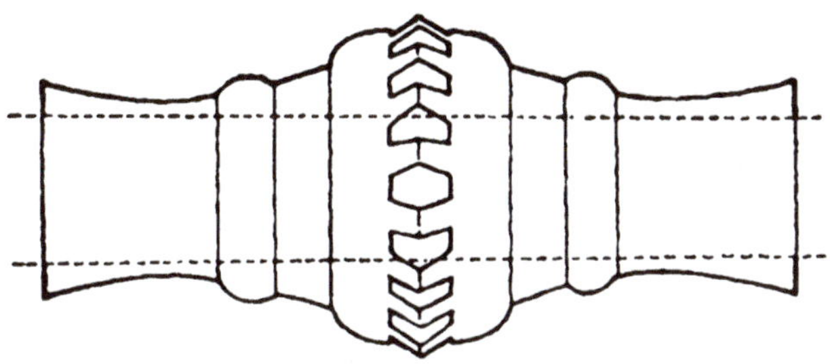

FIGURE 16B Structure of the wheel, consisting of rim, spokes and hub
Source: Dai Wusan, *Illustrated and Explained Kao Gong Ji*, Ji'nan: Shandong Pictorial Publishing House, 2003, p. 29, 101.

FIGURE 17
Bronze axle caps
Source: *Das Alte China – Menschen und Götter im Reich der Mitte* (Ancient Men and Gods in the Middle Kingdom), Hirmer Verlag München, 1995.

FIGURE 18
An axle's end pin
Source: Ibid.

The wheel's rim was composed of two, three or four bent, wooden parts. A tube-shaped bronze part was pushed over the joint. Using holes in the wheel rim segments and in the bronze part, the segments were fixed with leather strips. Fig. 19 shows the rubbing of a wall relief representing the manufacture of a wheel from several rim segments.

The next figure shows a wheel from a carriage made by traditional technology, exhibited in a museum of folk culture in Xining 西宁, Qinghai Province 青海.

In Fig. 20, an interesting detail can be seen: a decorative bronze ring is fixed onto the end of the hub. Such rings were used in order to attach a leather cover, as mentioned in the "Kao Gong Ji", and served to seal the space between the hub hole and axle, which was lubricated with fat. Moreover, the leather cover had a decorative function too, because it was carefully processed.

An important innovation in the carriages was the bent pole, which was more advantageous than a straight pole since it better transmitted the force of the horses to the carriage.

The invention of the breast strap harness in particular, which replaced the older throat-and-girth harness based on the bent pole, made transport using horse-drawn carriages much more effective, since the new harness system

FIGURE 19 Wall relief from the Han 汉 dynasty, with a representation of the manufacture of a wheel, excavated in Hongshan 洪山, Jiaxiang 嘉祥 (Shandong 山东 Province) Source: J. Needham, *Science and Civilisation in China*, Cambridge University Press, 1965, p. 81.

FIGURE 20 Wheel from a carriage made by traditional technology

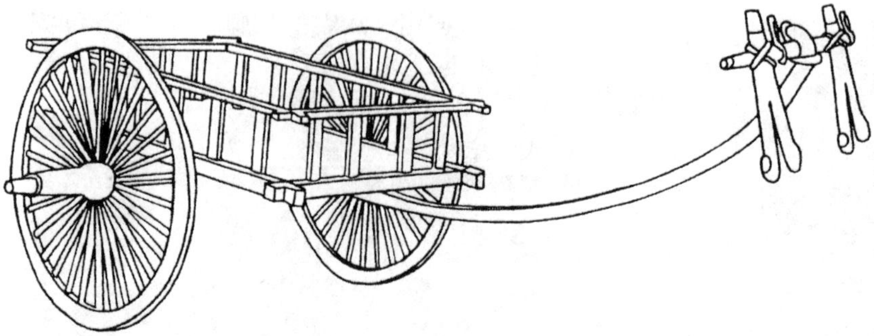

FIGURE 21 Carriage with bent pole

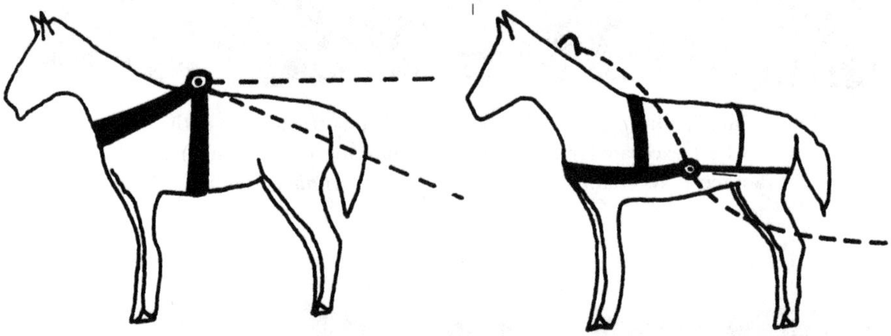

FIGURE 22 Two types of horse harnesses: left– old throat-and-girth harness, right – the newer breast strap harness using the bent pole
Source: *Imperial China – The Art of the Horse in Chinese History*, Kentucky Horse Park, 2000, p. 34.

enabled the horses to breathe more freely. Therefore, in the time of the creation of "Kao Gong Ji", more and more carriages were used that were drawn by only one horse.

Fig. 23 depicts the structure of the carriage's umbrella.

The carriage's umbrella, with its hub on the upper end of the supporting rod holding 28 spokes, is a delicate structure that embodies the high level of craftsmanship of this time.

In the tomb of Emperor Qin Shihuangdi 秦始皇帝 (259–210 BCE), an entire carriage with horses made from bronze was excavated, which represented the technique of horse-drawn carriages in detail.

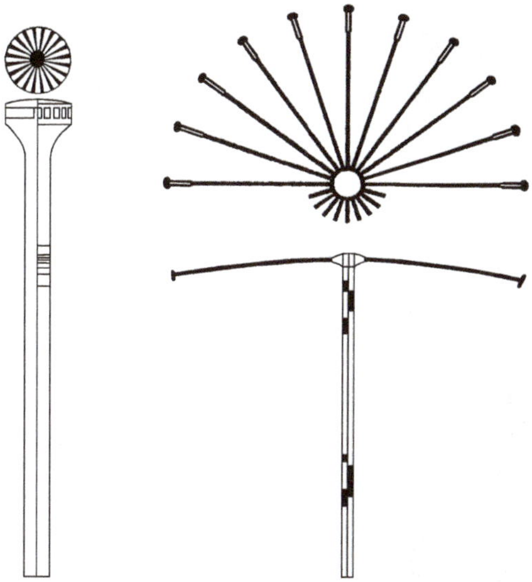

FIGURE 23
Components of the carriage's umbrella
Source: Wenren Jun, *Translated and Commented Kao Gong Ji*, Shanghai, Shanghai Ancient Books Press, 2008, pp. 30–31.

FIGURE 24 Bronze chariot from the tomb of Emperor Qin Shihuangdi
Source: *The Emperor Qin Shihuang and the Terracotta Sculptures of Warriors and Horses*. Shaanxi Culture Audio & Video Publishing House.

FIGURE 25
Bronze chariot from the tomb of Emperor Qin Shihuangdi (detail of the charioteer)
Source: Ibid.

2 Astronomy

In the chapter on the structure of carriages, it is said that the square carriage box symbolises the earth, and the round carriage umbrella the sky. The 30 spokes of the wheel symbolise the sun and the moon, and the 28 spokes of the carriage umbrella the stars. Here, the 28 lunar mansions *xiu* 宿 are meant, which represented certain constellations in ancient China. The 30 spokes of the wheels stand for the 30 days of the month.

The background to these symbols consists of the idea that heaven, earth and man are connected, which was applied here to the manufacture of carriages.

As early as in the Chunqiu 春秋 period, ancient Chinese astronomers established a spatial division of celestial bodies, which served to determine the beginning of the year and of the seasons. This spatial division represented the integration of fixed stars into constellations, which were located near the ecliptic. Altogether, 28 constellations were created, which were named "lunar mansions". By determining the position of the moon within the lunar mansions, the position of the sun relative to the stars was determined. The position of the sun indicated the beginning of the year and of the corresponding seasons.

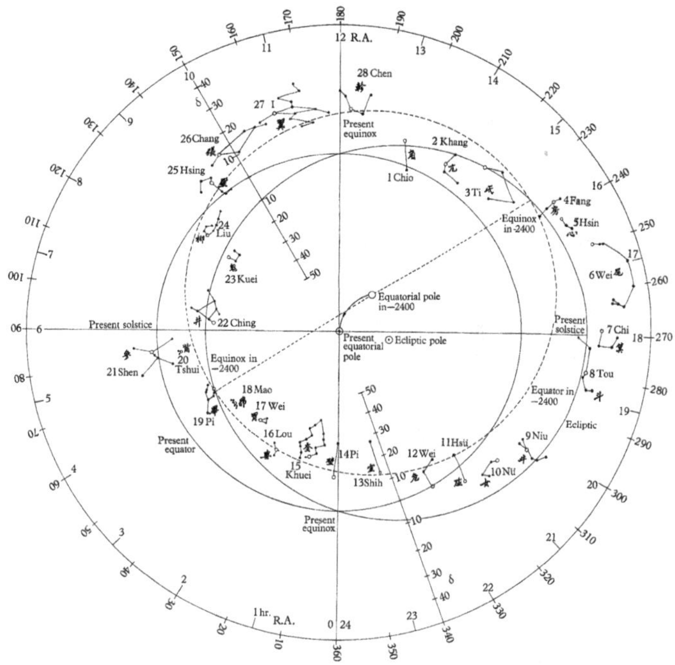

FIGURE 26 Star map with the star constellations represented by 28 lunar mansions. The map shows the ecliptic in its position today (solid line) and in its position around 2400 BCE (broken line). Source: J. Needham, *Science and Civilisation in China*, Cambridge University Press, 1959, vol. 3, opposite p. 250, Fig. 94.

The 28 lunar mansions were divided into four groups of 7 lunar mansions for each direction, as follows:

- In the east, summarised under the name *Canglong* 苍龙 (Green Dragon): *Jiao* 角, *Kang* 亢, *Di* 氐, *Fang* 房, *Xin* 心, *Wei* 尾, *Ji* 箕
- In the north, summarised under the name *Xuanwu* 玄武 (Dark Warrior): *Dou* 斗, *Niu* 牛, *Nü* 女, *Xu* 虚, *Wei* 危, *Shi* 室, *Bi* 壁
- In the west, summarised under the name *Baihu* 白虎 (White Tiger): *Kui* 奎, *Lou* 娄, *Wei* 胃, *Mao* 昴, *Bi* 毕, *Zi* 觜, *Shen* 参
- In the south, summarised under the name *Zhuque* 朱雀 (Red Bird): *Jing* 井, *Gui* 鬼, *Liu* 柳, *Xing* 星, *Zhang* 张, *Yi* 翼, *Zhen* 轸

Fig. 26 shows the arrangement of the lunar mansions. It is notable that the lunar mansions are relatively crowded along the ecliptic as it was located around 2400 BCE. This suggests that the determination of the lunar mansions happened very early.

As one can see on the star map in Fig. 26, the 28 lunar mansions are naturally not located on the ecliptic at equal distance intervals. However, they are well

FIGURE 27 Representation of the 28 lunar mansions on the lid of a lacquered box from the tomb of Marquis Yi of Zeng 曾侯乙 Source: R. Goepper (ed.), *Das alte China – Geschichte und Kultur des Reiches der Mitte* (Ancient China – History and culture of the Middle Kingdom), C. Bertelsmann Verlag München, 1988, p. 258.

suited for determining the beginning of the year and of the seasons through the aiming at a certain star of a lunar mansion.

Some of these lunar mansions are mentioned in "Kao Gong Ji", while their names changed in the course of time. A figurative representation of the 28 lunar mansions was found on a lid of a lacquered box in the tomb of Marquis Yi of Zeng 曾侯乙 (ca. 433 BCE) (see Fig. 27).

On the left side, one recognises the Green Dragon, which symbolises the lunar mansions of the east. On the right side, the White Tiger can be seen, which is the leader of the lunar mansions of the west. In the middle is the constellation of Big Dipper, which corresponds to the Great Bear. All the way around, characters indicate the 28 lunar mansions.

The lunar mansions mentioned in the chapter on carriages, which are related to the flag decoration of the war chariots, have the following form.

The lunar mansion Great Fire also is named Heart (*xin* 心) and represents the three stars at the top right. On the bottom left is the lunar mansion Tail (*wei* 尾), which consists of nine stars.

The lunar mansion Quail Fire is also named Willow (*liu* 柳) and is formed by the eight stars on the top right. Nearby, on the bottom left, is the lunar mansion Star (*xing* 星) (eight stars).

The lunar mansion Chopper (*fa* 伐) consists of four stars located on a line in the middle; it is framed by the six stars of the lunar mansion Shen 参.

FIGURE 28 Form of the lunar mansions Great Fire (*da huo* 大火) (left) and Quail Fire (*chun huo* 鹑火) (right)
Source: Dai Wusan, *Illustrated and Explained Kao Gong Ji*, Ji'nan: Shandong Pictorial Publishing House, 2003, p. 41.

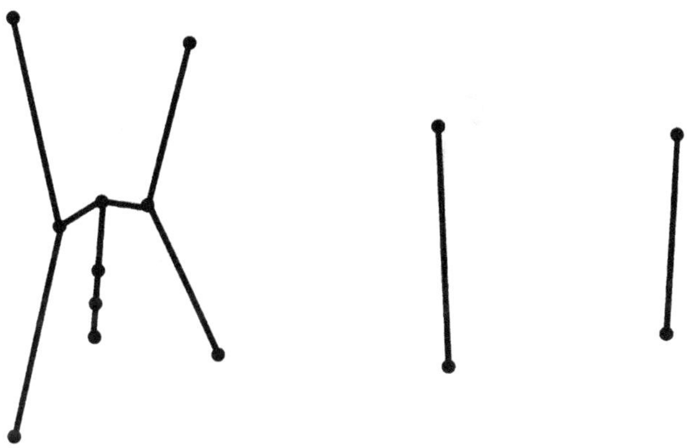

FIGURE 29 Form of the lunar mansions Chopper (*fa*) 伐 (left) and Chamber (*ying shi* 营室) (right)
Source: Dai Wusan 戴吾三, *Kao Gong Ji Tushuo* 考工记图说 (Illustrated and Explained Kao Gong Ji), Shandong huabao chubanshe 山东画报出版社, 2003, p. 41.

The lunar mansion Chamber (*ying shi* 营室) is located on the right side of the lunar mansion Wall (*bi* 壁). The lunar mansion Wall marks the spring equinox on the starry sky.

The lunar mansion Bow and Arrow is a part of the lunar mansion Well (*jing* 井).

Finally, in the chapter on the manufacture of carriages, the form of the pole is compared with the form of the lunar mansion *zhu* 咮, where *zhu* 咮

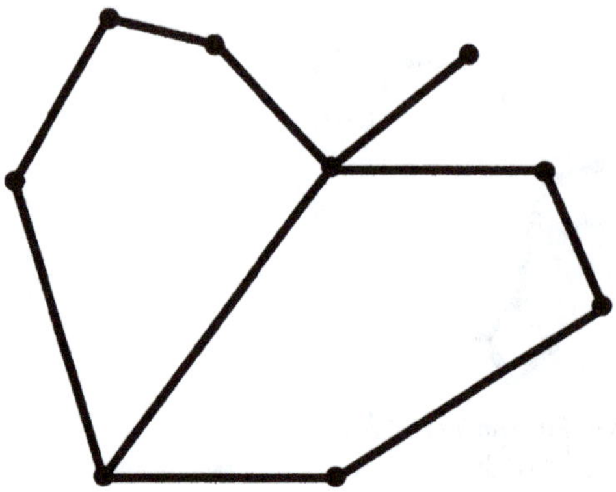

FIGURE 30
Form of the lunar mansion Bow and Arrow (*hu shi* 弧矢)
Source: Ibid.

FIGURE 31 Form of the pole of a carriage from the Zhanguo 战国 period, compared with the form of the lunar mansion Willow (*liu* 柳)
Source: Wenren Jun 闻人军, *Kao Gong Ji yizhu* 考工记译注 (Kao Gong Ji with translation).

is another name for the lunar mansion Willow (*liu* 柳). As is shown by the reconstruction of a carriage from the Zhanguo 战国 period, excavated near Linzi 临淄 in the province Shandong 山东, in Fig. 31, the form of the pole actually resembles the form of the lunar mansion Willow.

3 Bronze Technology

Six Mixing Ratios of Bronze Alloys

The six mixing ratios of bronze alloys intended for six different product groups, as described in the "Kao Gong Ji", represent the earliest written mention of bronze alloy compositions. Chinese and Japanese scientists have investigated this chapter most thoroughly.[3]

A big difficulty is posed by the fact that the term *"jin"* 金 used in these mixing ratios is ambiguous and, depending on the context, can mean metal, gold, copper or bronze. Nowadays, it is generally acknowledged that *"jin"* 金 means copper in this chapter. This was confirmed by a chemical analysis of archaeologically discovered bronze objects. Consequently, we get the following table of the six mixing ratios of bronze alloys.[4]

TABLE 4 Six mixing ratios of bronze alloys in "Kao Gong Ji"

Product group	Amount of copper, %	Amount of tin, %
Clocks and vessels	86	14
Cleavers and axes	83	17
Lances and halberds	80	20
Swords	75	25
Knives and arrowheads	71	29
Mirrors	67	33

Regarding the mixing ratio of the sixth product group "mirrors", the original wording states: "Copper, tin one half". This could be interpreted to say that one half of copper and one half of tin should be mixed. However, such an alloy would be too soft, preventing the mirror plane from being polished. Therefore, Zhou Shimin 周世民 interpreted this as follows: "Copper one, tin one half", which comes very close to the actual mixing ratio of bronze mirrors from the Chunqiu 春秋 and Zhanguo 战国 periods (770–221 BCE).[5]

3 Zhang Zigao 张子高, *Liu ji bie jie* 六齐别解 (A critical interpretation of the six mixing ratios), Qinghua daxue xuebao 清华大学学报, 1958, 6, pp. 159–165, He Tangkun 何堂坤, *"Liu ji" zhi guankui* 六齐之管窥 (Fragmentary examination of the six mixing ratios), Kejishi wenji 科技史文集, collection 15, Shanghai keji chubanshe 上海科技出版社, 1989, and Eikō Shimao 永康岛尾, *Chūgoku kagakushi* 中国化学史 (History of chemistry in China), Asakura Shoten 朝仓书店, 1995.

4 Zhou Shimin 周始民, *"Kao Gong Ji" liu ji chengfen de yanjiu* 考工记六齐成分的研究 (Investigation of the composition of the six mixing ratios in "Kao Gong Ji"), Huaxue tongbao 化学通报 54, 1978 3, pp. 182–185.

5 Ibid.

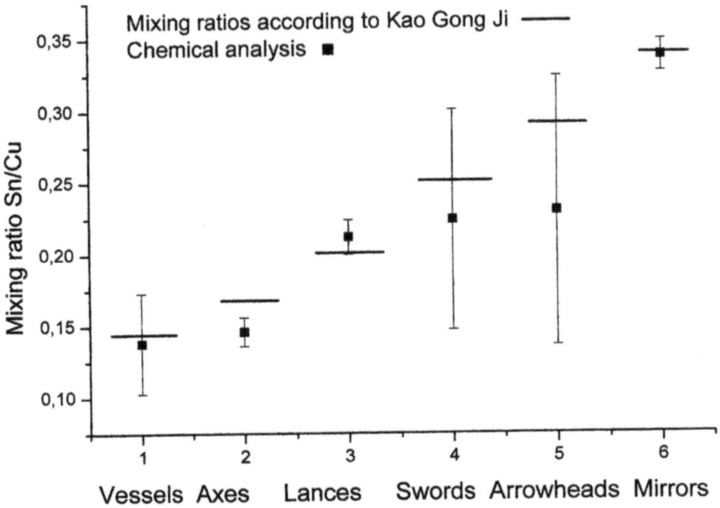

FIGURE 32 Mixing ratios of bronze alloys for the six product groups according to Table 4

The results of the chemical analysis of 77 archaeologically discovered bronze objects from the Chunqiu 春秋 and Zhanguo 战国 periods are depicted in Fig. 32.

In Fig. 32, the solid lines represent the mixing ratios provided in "Kao Gong Ji". This proves that, according to the chemical analysis, the actual mixing ratios in the first group ("vessels") and in the sixth group ("mirrors") correspond quite well to the mixing ratios provided in "Kao Gong Ji". Contrary to this, the results of chemical analyses of the remaining four product groups of axes, lances, swords and arrowheads do not correspond strongly to the mixing ratios in "Kao Gong Ji".

The classification of the products comprises:

- ritual objects (vessels and clocks)
- tools (axes)
- weapons (lances, swords, knives, arrowheads)
- consumer goods (plane and concave mirrors)

It is noticeable that products such as bronze components for carriages, agricultural tools and coins are not mentioned.[6]

6 Su Rongyu 苏荣誉, Hua Jueming 华觉明, Li Kemin 李克敏, Lu Benshan 卢本珊, *Zhongguo shanggu jinshu jishu* 中国上古金属技术 (Ancient Chinese Metallurgy), Shandong kexue jishu chubanshe 山东科学技术出版社, 1995.

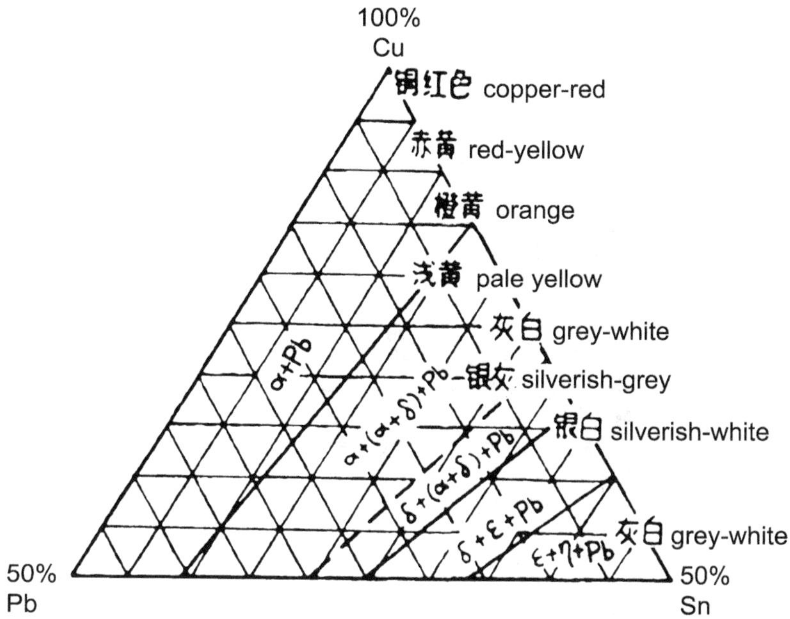

FIGURE 33 Ternary phase diagram of Cu-Sn-Pb alloys, with indication of the colours of the bronze alloys
Source: Zhou Shimin (周始民), *Research on the Six Mixing Ratios of Components in Kao Gong Ji*, Chemistry, 1978, 3, pp. 182–185.

The values for product groups 1 to 5 are single values, whereas the three values for the mirrors represent mean values for 100, 23 and 40 mirrors respectively.

Fig. 33 shows the ternary phase diagram between Cu, Sn and Pb. The colours of the corresponding Cu-Sn alloys are indicated on the line connecting Cu and Sn.

Fig. 33 shows that until 15% Sn content, the bronze alloys consist of the α-phase, and above 15% Sn until ca. 27% Sn, they consist of the (α+δ)-eutectoid. As Sn content increases, the red colour of pure copper changes to the typical pale yellow of bronze (for 20% Sn content) and then to grey-white respectively silverfish grey if the Sn content reaches ca. 30%.

Furthermore, Fig. 34 represents the relationship between the alloy contents and the strength of bronze alloys.

Fig. 34 makes clear that the highest tensile strength of bronze alloys (31 kp/mm^2) is achieved in bronze alloys with an Sn content from 5% to 20%. An increasing Pb content leads to a decreasing tensile strength of the alloy in every case.

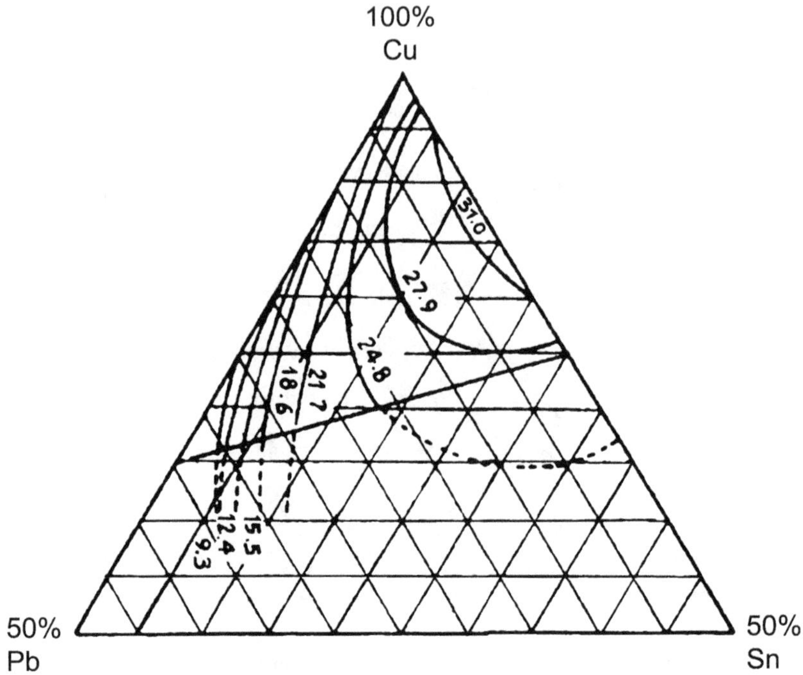

FIGURE 34 Ternary phase diagram of bronze alloys, indicating the relationship between Cu-, Sn- and Pb-content and the tensile strength (in kp/mm²) of the bronze alloys
Source: Ibid.

The casters of ancient China obviously possessed empirical knowledge about the achievable strength of different bronze alloys. Therefore, they produced a bronze alloy of the highest possible strength for vessels and axes. On the other hand, the different Sn contents described in "Kao Gong Ji" for the three types of weapons, which in principle required the same strength if they were to be used in battle, are not totally understandable. As the results of chemical analyses of corresponding artefacts prove, the mean values of these three types of weapons basically correspond to each other. However, one must also consider that the weapons for the rulers mainly served as showpieces. Therefore, it was tolerated that such weapons possessed insufficient strength (they were brittle). Furthermore, the material composition of swords was often not homogeneous, but rather sophisticated. The cutting edge of the sword was quite hard, but the rim was softer. In this way, the composition guaranteed the necessary toughness of the weapon.

Regarding the mirrors, a desired silver-like colour of the alloy was achieved because of its high Sn content (ca. 30%). Moreover, the brittleness of this alloy was advantageous, as it allowed for the polishing of the mirror's surface.

The "Kao Gong Ji" has described six mixing ratios for bronze alloys used in six different product groups, where the Sn content increases almost in equal steps from one product group to the next. Because the technical need for this equally increasing Sn content is not comprehensible, it seems to be that these mixing ratios were formulated with integer numbers, mainly to state simple relationships. It is possible that the casters of the bronze deviated from these simple relationships if the application of the products required this.

As an official book of the state Qi 齐, the "Kao Gong Ji" mostly describes handicrafts that served the luxury needs of the ruling class. This seems to be true for the casting of bronze objects as well. Therefore, bronze objects meant for the needs of broader society were manufactured according to mixing ratios that differed from the data given in the "Kao Gong Ji".[7]

Moreover, the deviations from the mixing ratios given in "Kao Gong Ji" can be explained by the following influences:

The "Kao Gong Ji" came into being in the state Qi 齐, mainly during the Zhanguo 战国 period. The stated values obtained through the chemical analyses of bronze artefacts are related to findings all over China, and also contain bronze discoveries from the late Shang 商 and the early Zhou 周 dynasties.

The investigation of the homogeneity of the Sn distribution in the famous vessel *mao gong ding* 毛公鼎 on six locations revealed that the Sn content ranged from 8.7% to 11.0%.[8]

The weighing of the copper and tin portions for the bronze alloy was not accurate, and could also lead to deviations from the rated mixing ratio.

Although a large amount of bronze artefacts has already been excavated, only a small amount of them underwent a chemical analysis. Therefore, the size of the statistical samples is so far too small to deliver proof of the scientific value of the six mixing ratios in "Kao Gong Ji".

The mixing ratios are also influenced by other alloy metals, like lead, iron and zinc, which depend on locally found ores. For instance, it was detected early on that bronze can be cast better if it contains 4 to 6% lead, because it will not deform during the process of cooling down.[9] Generally however, tin and lead could not be identified.

---

7 Su Rongyu 苏荣誉, Hua Jueming 华觉明, Li Kemin 李克敏, Lu Benshan 卢本山, Ancient Chinese Metallurgy, Ji'nan: Shandonmg Science & Technology Press, 1995.

8 Liu Guangding 刘广定, *Cong 'lu huo chun qing' tan 'liu ji'* 从炉火纯青谈六齐 (From the pure green-blue colour of the oven flames discuss the six mixing ratios), Science Monthly, 2008, 2, pp. 2–3.

9 Eikō Shimao 永康岛尾, *Chūgoku kagakushi* 中国化学史 (History of chemistry in China), Asakura Shoten 朝仓书店, 1995.

4 Bronze Objects – Vessels, Weapons, Mirrors

Ritual bronze vessels can be traced back to the Shang 商 dynasty. Based on inscriptions in several bronze vessels, they were clearly used for sacrificial aims. The following survey shows that there was a whole spectrum of typical bronze vessels.

Fig. 35 reveals that these ceremonial vessels were intended for drinking, storing beverages and food, and cooking (sacrificial) food. The three- and four-legged bronze vessels were used for cooking, that is, a fire could be lighted under the vessel.

It is noticeable that even the earliest bronze vessels excavated so far are shaped in a very artistic way and put high demands on bronze technology. The vessels were manufactured in assembled forms by the lost wax technology. Although ceramic vessels from the late neolithicum have been found that embody the original forms of some types of bronze vessels shown in Fig. 35, so far no bronze vessels are known that reflect the beginnings of bronze technology.

The heaviest bronze vessel for food sacrifices weighs 833 kg (see Fig. 36). For its casting around the mould, several smelting furnaces had probably been erected in order to melt the required large amount of bronze.

Bronze Weapons

In ancient China, five types of thrust weapons (*wu bing* 五兵) were distinguished, which are defined in different ways, depending on the source. The commentary to the book "Zhou Li" 周礼 (*Rites of the Zhou Dynasty*) by Zheng Xuan 郑玄 lists the following weapons: spear *ge* 戈, lance *shu* 殳, halberd *ji* 戟, lance *youmao* 酉矛 and lance *yimao* 夷矛. A commentary to the philosophical work Xun Zi 荀子 interprets the five weapons as follows: lance *mao* 矛, halberd *ji* 戟, axe *yue* 钺, shield *dun* 盾 and bow and arrow *gong shi* 弓矢. Besides descriptions of these weapons, the "Kao Gong Ji" also contains descriptions of the design of swords.

The sizes of these knives and the angle that they formed with the lance's shaft were prescribed so that they were as stable and as lethal as possible.

For the space between the sizes of the handle and the blade of a sword, fixed proportions were prescribed. Moreover, dependent on the rank of the dignitaries, the weights of the swords were laid down according to the principle "the higher the rank of the carrier, the heavier the sword."

Knives and arrowheads possessed a higher tin content, which made the material harder (but also more brittle).

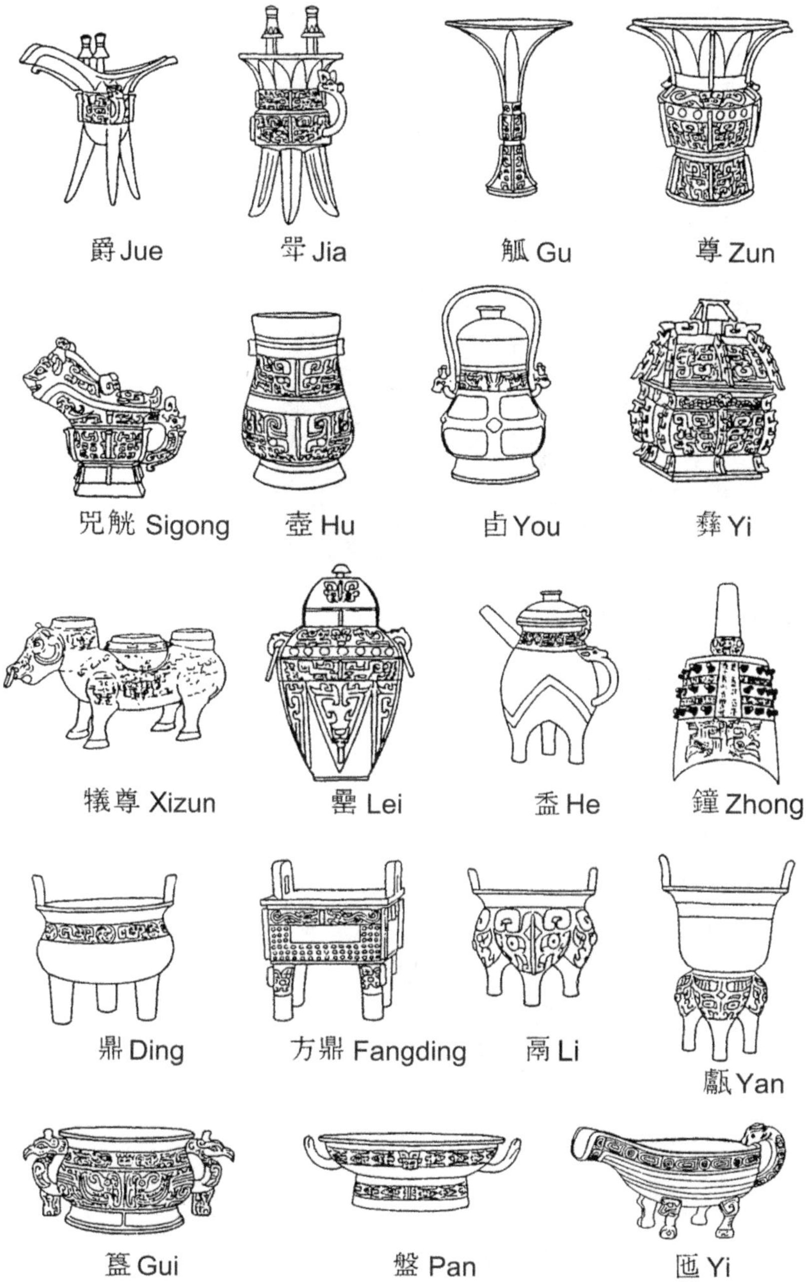

FIGURE 35 Types of bronze ceremonial vessels
Source: E. Shimao 永康岛尾, *Chūgoku kagaku shi* 中国化学史 (History of chemistry in China), Asakura Shoten 朝仓书店, 1995, p. 69.

FIGURE 36 Rectangular bronze vessel for food sacrifices of type *ding* 鼎 (Shang 商 dynasty). Weight 833 kg, length 110 cm, height 133 cm, width 79 cm
Source: *Exhibition of Chinese History*, National Museum of Chinese History, 1998, p. 37.

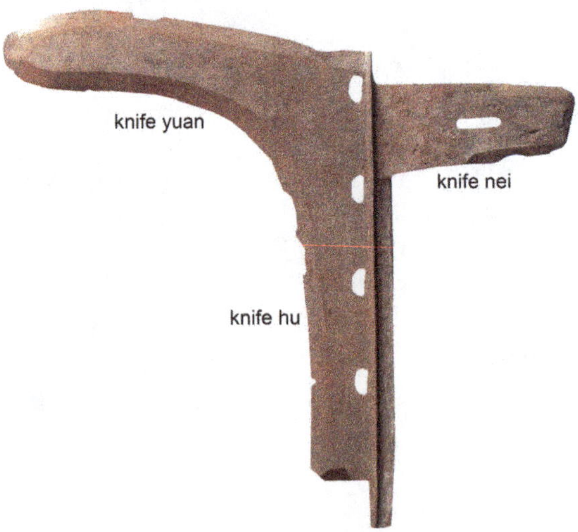

FIGURE 37 Lü Buwei's 吕布韦 lance, with the knives yuan 援, hu 胡 and nei 内 (Zhanguo 战国 period)
Source: *Exhibition of Chinese History*, National Museum of Chinese History, 1998, p. 37.

FIGURE 38 Battleaxe

FIGURE 39 Bronze swords (Zhanguo 战国 period)
Source: www.Baidu.com, article 战国青铜剑 (bronze swords of the Zhanguo period), searched on 22.03.2018.

FIGURE 40
Bronze axe *yue* 钺 (late Shang 晚商 dynasty, ca. 1300–1030 BCE)
Source: *Das alte China – Menschen und Götter im Reich der Mitte* (Ancient China – Men and gods in the Middle Kingdom), Hirmer München, 1995, p. 235.

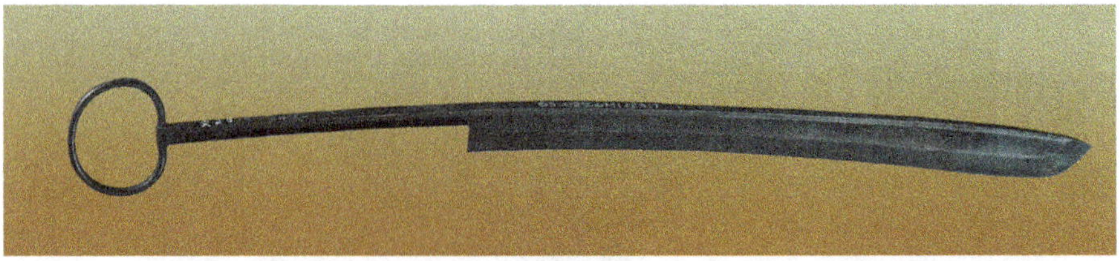

FIGURE 41 Erasing knife for correcting script on bamboo strips (Zhanguo 战国 period)
Source: *Das alte China – Menschen und Götter im Reich der Mitte* (Ancient China – Men and gods in the Middle Kingdom), Hirmer München, 1995, p. 315.

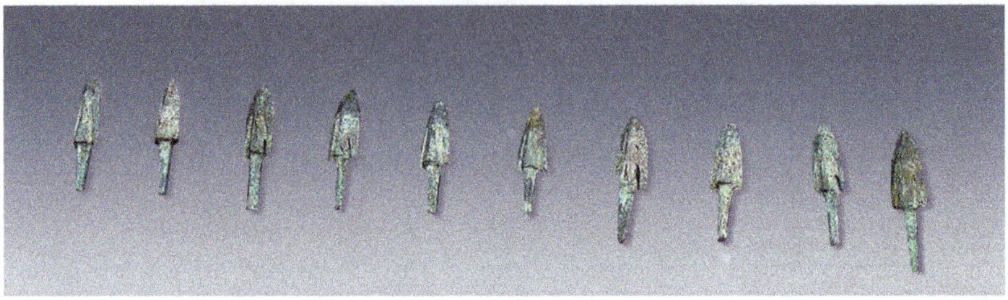

FIGURE 42 Bronze arrowheads from the Chunqiu 春秋 period
Museum of Shaanxi 陕西 province in Xi'an 西安

The erasing knives are bent with a central angle of 60°, so that six erasing knives form a circle.

Bronze Plane and Concave Mirrors

It is noteworthy that the "Kao Gong Ji" mentions plane and concave mirrors, in connection with the six mixing ratios for bronze alloys, as the sixth product group.

The plane bronze mirrors in ancient China were smoothly polished on the front side, so that one could observe the mirror's image. On the back is an eye to hang the mirror. Moreover, the back is richly decorated as a rule.

Concave mirrors were used as burning mirrors by focusing the sunlight to generate fire. This effect is mentioned first in the books "Mojing" 墨经 and "Zhou Li" 周礼 (chapter Qiu Guan 秋官), and later in the book "Huai Nan Zi" 淮南子 (ca. 120 BCE). The book "Mojing" 墨经, which came into being around the same time as the "Kao Gong Ji", already mentions the idea of the focal point on which the light beams are focused.

Fig. 44 shows a concave mirror from the Chunqiu 春秋 period, which was excavated in Shangcunling 上村岭 near the city Sanmenxia 三门峡. This finding proves that as early as this period, concave mirrors' ability to focus light beams was known.

FIGURE 43 Bronze mirrors from the Zhanguo 战国 period
Museum of Shaanxi 陕西 province in Xi'an 西安

FIGURE 44
Concave mirror from the Chunqiu 春秋 period
Source: www.Baidu.com, article 春秋凹形镜 (concave mirror of the Chunqiu period), searched on 22.03.2018.

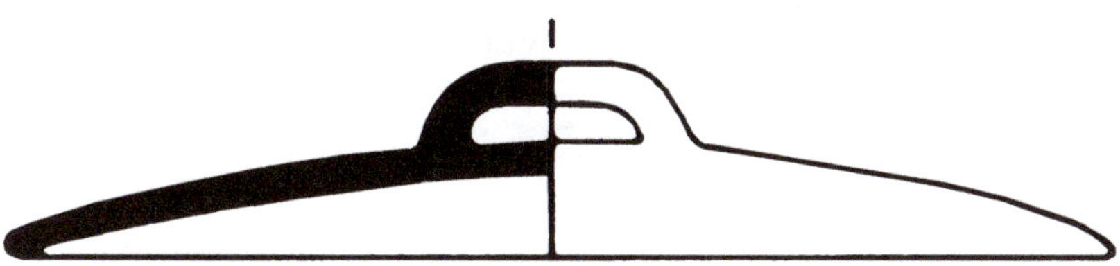

FIGURE 45 Cross section of the concave mirror in Fig. 44
Source: *Dai Wusan, Kao Gong Ji Tu Shuo*, Shandong Huabao Chubanshe, 2003, p. 43

5 Metrology

Several scientists have found that the system of measurement units used in "Kao Gong Ji" derived from the state Qi 齐. Therefore – and due to other facts – they concluded that this work originated in the state Qi.

The system of measurement units in ancient China comprised measurement units for length, mass and volume.

The basic unit of length is the Chi 尺 (foot). From an investigation by Wenren Jun 闻人军,[10] it was concluded that the Chi 尺 corresponded to 19.65 cm.

Downwards, the Chi 尺 was divided decimally:

> 1 Chi 尺 = 10 Cun 寸; 1 Cun 寸 = 10 Fen 分 (the unit Fen also is called Mei 枚).

The following units were larger than the Chi:

> 1 Xun 寻 = 8 Chi 尺; 1 Chang 常 = 2 Xun 寻; 1 Ke 柯 = 3 Chi 尺.

The "Kao Gong Ji" mentions that in specific fields, for instance in civil engineering, special units were used, but that these were connected to the universal units of length. Within rooms of houses, a small table *ji* 几 (3 Chi = 59 cm) was used as unit of length; in the halls, the bamboo mat *yan* 筵 (9 Chi = 177 cm); in the palace, the (regular) unit of length *xun* 寻 (8 Chi = 157 cm); outdoors, the step *bu* 步 (5 Chi = 98 cm); and for the streets, the width of the wheel track *gui* 轨 (8 Chi = 157 cm) was used. 300 *bu* 步 gives 1 Li 里 (471 m).

For the manufacture of carriages, the length of an axe grip Ke 柯 (= 3 Chi = 59.1 cm) was used as a unit of length.

The basic unit of mass is the Jin 斤 (literally: axe). In the region of the former state Qi, weights were found whose mass amounted to 198.4 g, which possibly equals a Jin 斤. Furthermore, the mass unit Jun 钧 was used, where 1 Jun 钧 = 30 Jin 斤. Finally, the still larger unit Shi 石 was used.

Furthermore, for mass, the following mass units were valid: Huan 锾 (5.95 g), where 1 Huan = 11 13/25 Zhu 铢 (1 Zhu = 0.6 g). Lüe 锊 (82.7 g), where 1 Lüe = 6 2/3 Liang 两 (1 Liang = 12.4 g). The weight of mass units Mou 侔 and Di 邸, which are mentioned in "Kao Gong Ji", is unknown.

The basic unit of volume was the Sheng 升, which corresponded to ca. 205 ml.

---

10 Wenren Jun 闻人军, *"Kao Gong Ji" Qi chi kao bian* "考工记" 齐尺考辨 (Investigation of the measurement unit Chi from state Qi in Kao Gong Ji), Kaogu 考古, 1983, 1, pp. 61–65.

Besides this, the following volume units were used. They were subdivided mainly into steps of four, according to an old system of measurement units:

> 1 Zhong 钟 = 10 Fu 釜, 1 Fu 釜 = 4 Ou 区, 1 = Ou 区 = 4 Dou 豆, 1 Dou 豆 = 4 Sheng 升.

Later, units of volume in the state Qi 齐 developed a system of subdivision mainly with five units using mainly five steps:

> 1 Zhong 钟 = 10 Fu 釜, 1 Fu 釜 = 5 Ou 区, 1 Ou 区 = 5 Dou 豆, 1 Dou 豆 = 5 Sheng 升.

Furthermore, the volume unit Yu 斞 is mentioned, which possibly amounted to 3.6 ml.

The size of the volume units in the state Qi 齐 could be reconstructed using several volume standards, the most famous among which are the volume standards Youli 右里 and Zihezi 子禾子.

The standard Youli has a volume of 1025 ml, so that it embodies 5 Sheng 升.

FIGURE 46 Bronze volume standard Youli 右里 from the state Qi 齐
Source: Qiu Guangming 丘光明: *Zhongguo wulixue shi daxi – jiliang shi* 中国物理学史大系-计量史 (Series of books on history of physics in China – A history of metrology), Hunan jiaoyu chubanshe 湖南教育出版社 (2002) 98–111.

FIGURE 47 Bronze volume standard Zihezi 子禾子 for the measurement unit Fu 釜 from the state Qi 齐
Source: Qiu Guangming (丘光明), *Series of Books on History of Physics in China – A History of Metrology*, Changsha: Hunan Education Publishing House, 2002, p. 98–111.

Using an inscription, the standard Zihezi was dated to the time 404–388 BCE. This cauldron was commissioned by senior official Zi Hezi 子禾子, who later became the ruler of Qi 齐. It has a volume of 20,460 ml, which corresponded to one Fu 釜. The inscription on the standard states that a strict system of metrology existed in the state Qi.

The volume of the standards Youli 右里 and Zihezi 子禾子 had a complicated form, which at that time could not be calculated yet. Therefore, the volume was obviously determined by adjustment of the vessel's height during their manufacture.

Finally, in the late Zhanguo 战国 period, a decimal system of volume units gained acceptance: 10 Sheng 升 = 1 Dou 豆, 10 Dou 豆 = 1 Fu 釜.[11]

Whereas we have already seen that in one state, Qi 齐, an old and a new system of volume units existed in parallel, the situation became still more

11 Ibid.

complicated in the China of that time because in each state, measurement units with different designations and values were valid.

The "Kao Gong Ji" describes a special system of angle measures, which is cut to the needs of workmen. Starting from the right angle 90° "ju" 矩, the following angles were defined:

1 xuan 宣 = 1/2 × 90° = 45°
1 zhu 欘 = 45° + 1/2 × 45° = 67.5°
1 ke 柯 = 67.5° + 1/2 × 67.5° = 101.25°
1 qingzhe 磬折 = 101.25° + 1/2 × 101.25° = 151.875°

Moreover, there were terms for pointed angles (已句) and for obtuse angles (已倨).

A universal angle measure system did not exist in ancient China, contrary to ancient Babylon, where for the full circle an angle of 360° was defined. However, China developed the system of angles outlined above, which was based on the right angle and the bisection of angles derived from it.[12] It proved to be very practical for the demands of workmen. With the development of astronomy as early as in the Zhanguo 战国 period, the coordinates of stars in star tables are given in angle degrees. The full circle was hereby defined as 365.25°; this means that the journey of the earth around the sun during one day corresponded to 1°.

The "Kao Gong Ji" moreover contains a hint which seems to show that the concept of the full circle existed. In the chapter on the manufacture of bows, the length of the bows is expressed by using parts of the circumference of the full circle. For instance, it is stated that the length of the bow for a vassal duke should be 1/7 of the circumference.

It is interesting that the name of the last angle, "qingzhe" 磬折, is connected to the stone chime, because it means "fracture side of the stone chime". Indeed, the stone chimes, which originate from the time of "Kao Gong Ji", have an angle between the two upper sides that is nearly 152°. But in "Kao Gong Ji", the upper angle of the stone chime is defined as 90° + 1/2 × 90° = 135°. Archaeological findings of stone chimes show this angle as well. More details on this are provided in the commentary on musical instruments.

Very remarkable is the combined measurement standard Li Shi Liang 栗氏量, for volume, length, mass and ton frequency, described in "Kao Gong Ji".

This measurement standard consists of a large container in the middle and two small containers that are attached like ears on the left and the right side.

12 Guan Zengjian 关增建, *"Kao Gong Ji" jiaodu gainian chuyi* "考工记" 角度概念刍议 (A humble opinion on the concept of angle in "Kao Gong Ji"), Ziran bianzhengfa tongxun 自然辩证法通讯, 2000, 2, pp. 72–76.

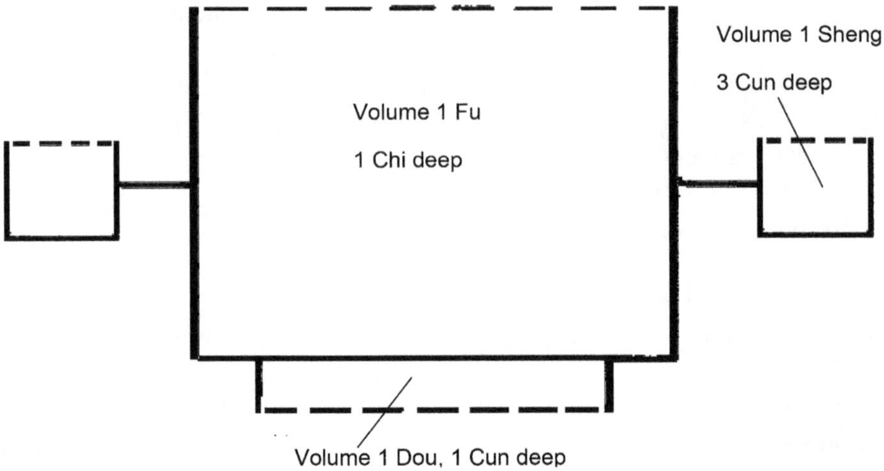

FIGURE 48 Structure of the standard Li Shi Liang 栗氏量 of "Kao Gong Ji"

The container in the middle on the upper side represents the volume unit 1 Fu 釜 and on the lower side 1 Dou 豆. The "ear containers" attached on the sides each represent 1 Sheng 升. The container for Fu 釜 is inside round and has an inscribed square with an edge length of 1 Chi 尺 and a depth of 1 Chi 尺. The lower container, for the volume unit Dou 豆, has a depth of 1 Cun 寸. The "ear container" for the volume unit Sheng 升 has a depth of 3 Cun 寸. In this way, the standard Li Shi Liang 栗氏量 embodies the three volume units Fu 釜, Dou 豆 and Sheng 升. The depth of the container for the volume unit Fu 釜 of 1 Chi 尺 serves as a measure of length. The mass of the entire combined measurement standard of 1 Jun 钧 serves as mass standard. Furthermore, it is indicated that if the container is struck (possibly in a certain location, which is not described in "Kao Gong Ji"), it emits the tone Gong 宫 of the standard pipe Huangzhong 黄钟. Therefore, this measurement standard also represents a standard of tone frequency, which was significant for the stability of court music.

This reconstruction of the standard Li Shi Liang 栗氏量 follows the ideas of Dai Zhen 戴震 and Wu Chengluo 吴承洛.[13] The reconstruction is based on the assumption that the old system of volume units is applied, namely:

1 Zhong 钟 = 10 Fu 釜
1 Fu 釜 = 4 Ou 区
1 Ou 区 = 4 Dou 斗
1 Dou 斗 = 4 Sheng 升

13 Guan Zengjian 关增建, *Xian Qin biaozhun liangqi* 先秦标准量器 (Measurement standards of the pre-Qin period), unpublished manuscript.

Since according to this relationship 1 Dou 斗 = 1/16 Fu 釜, the depth of the Fu-container is 1 Chi 尺 and the depth of the Dou-container is 1 Cun 寸 (where 1 Cun 寸 = 1/10 Chi 尺), the outer diameter of the Dou-container must be smaller than that of the Fu-container, as shown in Fig. 48.

However, it is also possible that the standard Li Shi Liang 栗氏量 embodies the system of volume units that was introduced later, where the relationship 1 Dou 斗 = 1/10 Fu 釜 was used. In this case, the diameters for the Fu- and the Dou-container are the same. This reconstruction of the standard Li Shi Liang 栗氏量 was proposed by researchers Chen Mengjia 陈梦家 and Qiu Guangming 丘光明.

This design of the standard seems to be more likely, because the combined measurement standard Jia Liang 嘉量 (see Fig. 50) from a later period, which still exists, has the same structure. Therefore, the standard Li Shi Liang 栗氏量 could have had a design as presented in Fig. 49.

The "Kao Gong Ji" mentions that the device Li Shi Liang was not used to impose taxes, but exclusively as a measurement standard. Therefore, this device can be regarded as the first measurement standard in the world.

The manufacture of a measurement standard cast from bronze testifies to the precision of bronze casting in the Chunqiu 春秋 and Zhanguo 战国 periods. The manufacture of the bronze alloy is described as follows. First, copper and tin are melted and refined. Then, the amount of copper and tin necessary for the bronze alloy is determined through weighing. Next, the volume of the bronze is determined by pouring water into the mould and weighing the water.

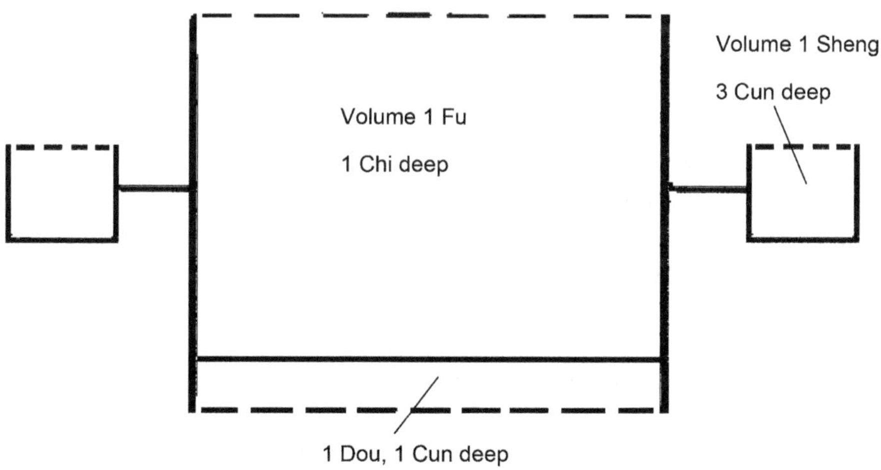

FIGURE 49 Reconstruction of the standard Li Shi Liang 栗氏量 according to the proposal of researchers Chen Mengjia 陈梦家 and Qiu Guangming 丘光明.[5]

FIGURE 50 Bronze volume standard for the unit Sheng 升, manufactured at the order of Shang Yang 商鞅 in the year 344 BCE. The original is kept in the museum of Shanghai 上海.

The text does not explain whether the density of bronze was known. Possibly, this procedure determined the volume of the to-be-cast liquid bronze.

From the Zhanguo 战国 period, another famous volume standard is passed on, namely Shang Yang's 商鞅 bronze volume standard for the unit Sheng 升. Shang Yang 商鞅 (?–338 BCE) was a minister and reformer in the state Qin 秦. According to the inscription on this standard (see Fig. 50), in the year 344 BCE. Shang Yang 商鞅 ordered the manufacture of this volume standard with a rectangular shape, where 1 Sheng 升 is defined as 16 1/5 cubic Cun 寸. By defining the volume through the geometrical sizes of the vessel, he created a calculable volume standard. The same approach can be observed in the structure of the combined measurement standard Li Shi Liang 栗氏量.

The sizes of this standard were measured as follows: length = 124.774 mm, width = 69.742 mm, depth = 23.23 mm. For the value of 1 Cun 寸 = 2,32 mm, one obtains the volume of this standard in units of Cun 寸: length 5.4 Cun 寸, width 3 Cun 寸, depth 1 Cun 寸, where $5.4 \times 3 \times 1 = 16.2$ cubic-Cun 寸.[14]

The reign of Wang Mang 王莽 (8–23 CE) created the combined measurement standard Jia Liang 嘉量, which now is kept in the Palace Museum in Taibei 台北 (see Fig. 51). This combined measurement standard expresses the high status that metrology had for the administration of the state.

14 Qiu Guangming, *Series of Books on History of Physics in China – A History of Metrology*, Changsha: Hu'nan Education Publishing House, 2002, p. 179.

FIGURE 51 Combined measurement standard Jia Liang 嘉量 from the reign of Wang Mang 王莽

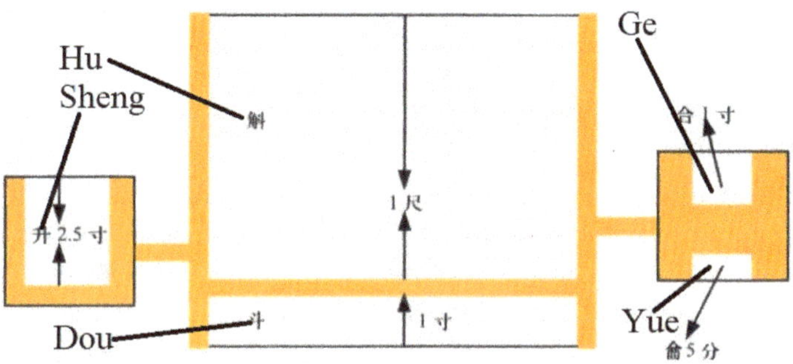

FIGURE 52 Cross section through the combined measurement standard Jia Liang 嘉量
Source: *Explanation to the combined measurement standard Jia Liang*, Measurement Center Xinzhu, Taiwan, 2004

As can be seen in Fig. 51, this combined measurement standard embodies the five volume units Hu 斛, Dou 斗, Sheng 升, Ge 合 and Yue 龠.[15]

In Europe, a comparable combined measurement standard is known to have been developed by Johannes Kepler in Ulm in the year 1617. He created

15 Guan Zengjian 关增建: *Zhongguo jiliang shi de guibao – xin mang tong jia liang* 中国计量史的瑰宝-新莽铜嘉量 (Treasure of Chinese metrology history – the bronze measurement standard Jia Liang from the rule of Wang Mang), in: Jiliang shihua 计量史话 (Talks on metrology history), China Metrology Publishing House, 2010, 81–85.

FIGURE 53
Combined measurement standard "Kepler vessel" for the measurement units of length, volume and mass
Source: K. Herrmann, *A comparison on the development of metrology in China and the West*, Wirtschaftsverlag NW, 2009, p. 91.

the "Kepler vessel", which embodied the measurement units for length, volume and mass.

6 Musical Instruments

The book "Kao Gong Ji" describes bells and sets of bells, chime stones and sets of chime stones and drums as musical instruments, which were of special importance to court music. The stands for clock and stone chimes were a peculiarity, executed in a very elaborated way according to laid down aesthetic principles. Besides the manufacture of these musical instruments, the knowledge described in "Kao Gong Ji" is of special interest for understanding how the pitch of these instruments could be influenced.

Bells and Sets of Bells

The following figure displays the structure of old Chinese bells, which were hung on a stand as a set of bells, mostly in different sizes.
The "Kao Gong Ji" has a detailed list of terms for the particular parts and regions of bells, which are presented in Fig. 54.

On sets of bells, one distinguishes between two different kinds of hanging: vertically hung bells (as shown in Fig. 56) and bells hung at a slant. The last were hung on the ring *xuan* 旋 (see Fig. 54).

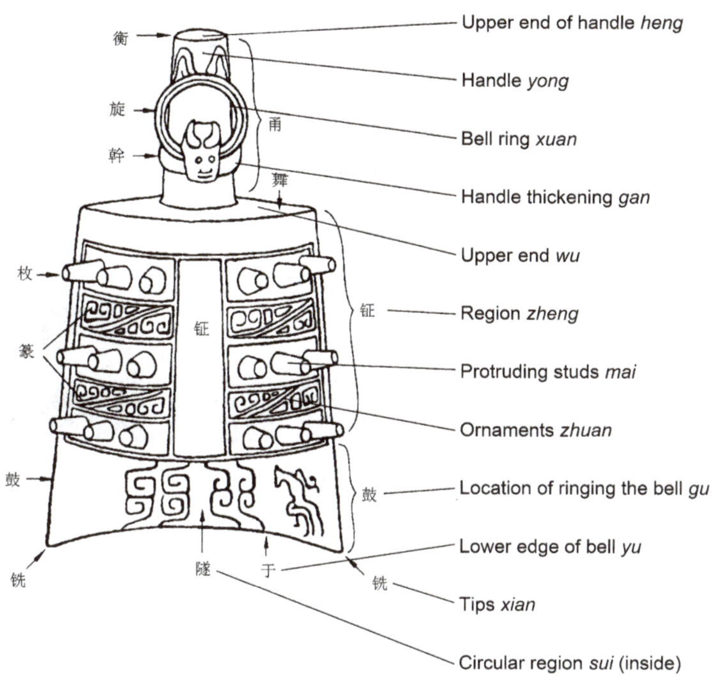

FIGURE 54 Structure of an old Chinese bronze bell
Source: Dai Wusan 戴吾三, *Illustrated and Explained Kao Gong Ji*, Shandong Pictorial Publishing House, 2003, p. 47.

FIGURE 55
Bronze bell (early Zhanguo 战国 period)
Shanghai 上海 Museum

FIGURE 56 Set of bells from the Marquis Su of Jin 晋侯苏 (Western Zhou 西周 dynasty, mid-9th century BCE)
Shanghai Museum

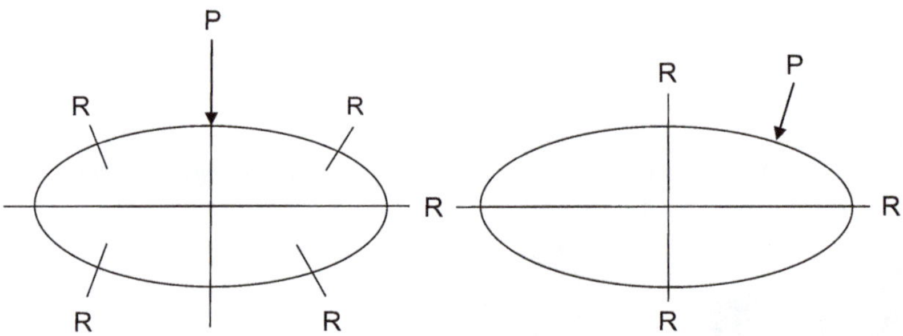

FIGURE 57 Locations for hitting a bell (P – location for hitting the bell, R – wave knot)
Source: Dai Wusan, Illustrated and Explained Kao Gong Ji, Ji'nan: Shandong Pictorial Publishing House, 2003, p. 47.

It is remarkable that the shape of the bells is defined on the basis of a modularised structure, as we already have seen for the structure of carriages. Here, the length of the opening of the bell *xian* 铣 is the basic measure from which the other measures are derived.

A peculiarity of the bells belonging to sets of bells consists of the fact that they have an elliptic cross section. Due to this, each bell could generate two different tones, depending on where the bell was hit (see Fig. 56).

The bells developed from bronze sleighbells where earliest findings of sleighbells are 4,000 years old.

The "Kao Gong Ji" explains, clearly based on empirical knowledge, how one can influence the sound of the bell. The wall's thickness and the size of the bell's opening and body were identified as factors that influenced its sound. As a final step, the sound was corrected by filing off material on the inner side of the bell, a process of which corresponding processing traces can be found on bells.

The stands for sets of bells were manufactured by a specialised carpenter, *zi ren* 梓人, and were decorated with animal statues.

Interestingly enough, in this section, the "Kao Gong Ji" provides a classification of large animals that today are referred to as vertebrates:

- Animals with horns
- Animals without horns
- Men
- Feathered animals
- Scaly animals

This classification also shows that men were understood to be a part of the fauna.

From the point of view of aesthetics, the "Kao Gong Ji" states that statues of men who carry the cross-bars fit well with sets of bells, with their deep, piercing tones. On the other hand, statues of birds were preferred as carriers of the cross-bars of sets of stone chimes, a better fit with the pure tones of chime stones.[16] Figs. 58 and 59 each show a set of bells and the statue of a bird as carrier for a set of chime stones. These artefacts were found in the tomb of the Marquis Yi of Zeng 曾侯乙 and correspond to the descriptions of "Kao Gong Ji".

Chime Stones and Sets of Chime Stones

Chime stones are a percussion instrument typical to classical Chinese music. The chime stones have a shape as shown in Fig. 60. They are hung on a stand with the hole above. Most often, the chime stones form a set of five or more stones.

Fig. 61 displays a set of chime stones in which chime stones of graduating sizes are hung in two rows.

The obtuse angle on the upper side of the chime stone amounts to 135° according to the data in "Kao Gong Ji". Here, a special angle measuring system becomes evident, designed for the needs of workmen. Starting from the right

16 Liu Dunyuan 刘敦愿, *Translation of the Chapter "The Carpenter Zi Ren Manufactures Stands for Sets of Bells" in Kao Gong Ji And the Observed Art Theory for the Carved Ornaments*, Art Research, 1985, 2, pp. 60–68.

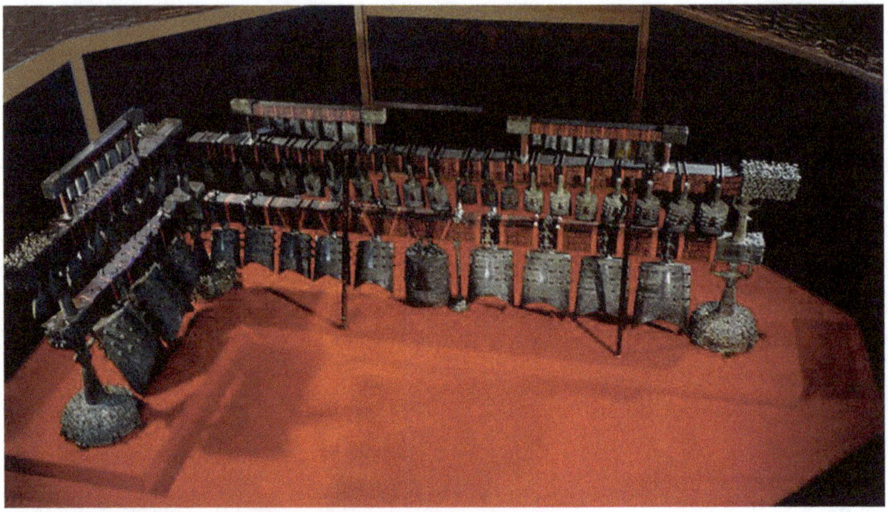

FIGURE 58 Set of bells from the tomb of Marquis Yi of Zeng 曾侯乙

FIGURE 59
Statue of a bird as carrier of the cross-bar in a set of chime stones from the tomb of Marquis Yi of Zeng 曾侯乙

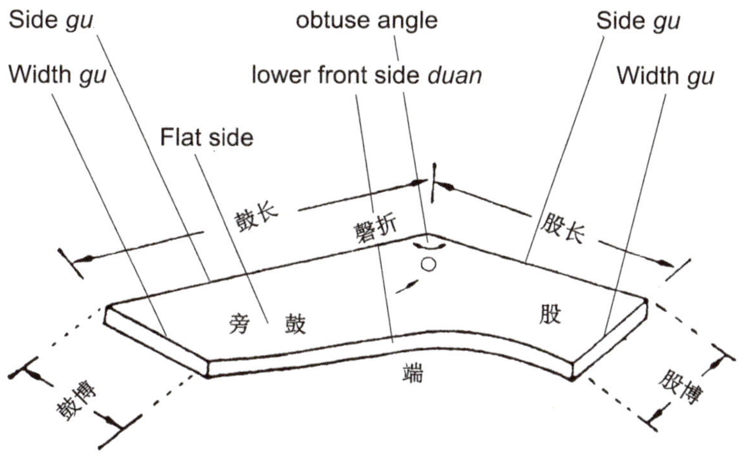

FIGURE 60 Geometry of a chime stone "*qing*" 磬 according to the description in "Kao Gong Ji"
Source: Dai Wusan, *Illustrated and Explained Kao Gong Ji*, Ji'nan: Shandong Pictorial Publishing House, 2003, p. 65.

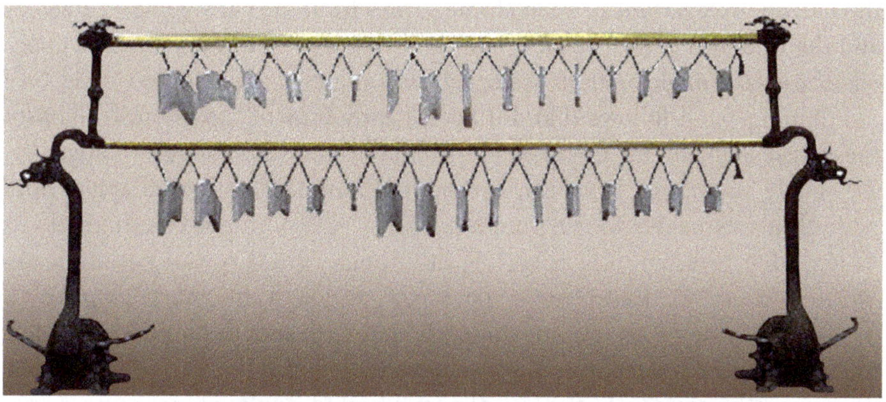

FIGURE 61 Set of chime stones from the tomb of Marquis Yi of Zeng 曾侯乙

angle of 90°, "*ju*" 矩, the following angles are defined (see also commentary on metrology):

1 *xuan* 宣 = 1/2 × 90° = 45°
1 *shu* 欘 = 45° + 1/2 × 45° = 67.5°
1 *ke* 柯 = 67.5° + 1/2 × 67.5° = 101.25°
1 *qingzhe* 磬折 = 101.25° + 1/2 × 101.25° = 151.875°

It is interesting that the name of the last angle, "*qingzhe*" 磬折, is connected with the chime stones, because it means "fracture side of the chime stone". On

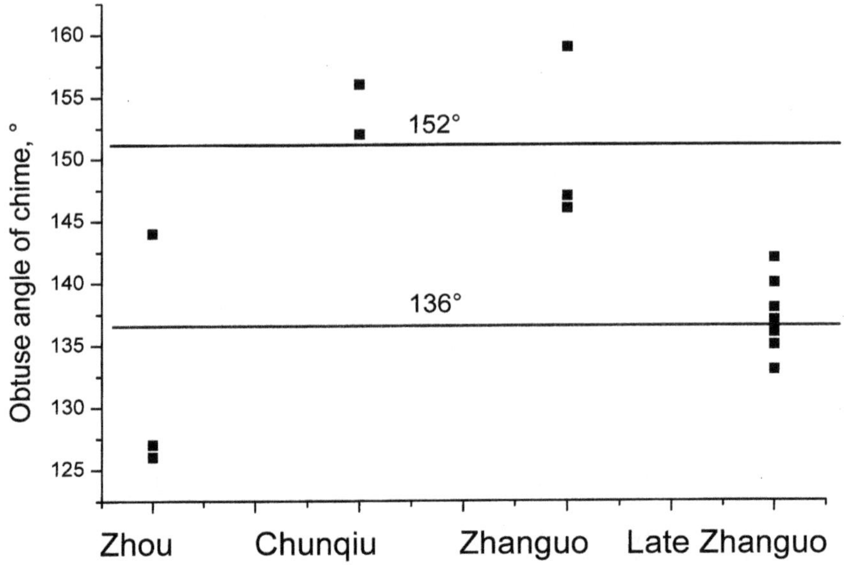

FIGURE 62 Change of the obtuse angle on the upper side of chime stones over time

the basis of excavations of such chime stones from the early Zhou 周 dynasty and the Chunqiu 春秋 (770–476 BCE) and Zhanguo 战国 (475–221 BCE) periods, the obtuse angle on their upper side was analysed as presented in Fig. 62.[17]

From Fig. 62, it follows that during the early time of the Zhou 周 dynasty, the obtuse angle on the upper side of chime stones was near 135° (= angle "*ju*" 矩 + angle "*xuan*" 宣). However, during the Chunqiu 春秋 and Zhanguo 战国 periods, it was nearer to 152° (angle "*qingzhe*" 磬折). In the late Zhanguo 战国 period, it reverted to near 135°. For this angle, the "Kao Gong Ji" indicates "*ju*" 矩 + "*xuan*" 宣, that is 90° + 45° = 135°, although the name "*qingzhe*" 磬折 (fracture side of the chime stone), which corresponds to 151.875°, points to chime stones. Indeed, this angle was realised during the Chunqiu 春秋 and Zhanguo 战国 periods. In the late Zhanguo period, angles at 135° appear once more. It is assumed that these angles were easier to manufacture and that the knowledge about the meaning of the term "*qingzhe*" 磬折 was lost.[18]

The geometry of the chime stones follows a modularised structure, where the side *gu* 股 is taken as the base element that the remaining elements of the chime stone are derived from. In this way, similar chime stones could be designed for an entire set.

17 Wenren Jun 闻人军, *Origin and Evolution of "Qing Zhe"*, "磬折" 的起源与演变, Journal of Hangzhou University (Natural Science Edition), 1986, 2, pp. 166–174.

18 Ibid.

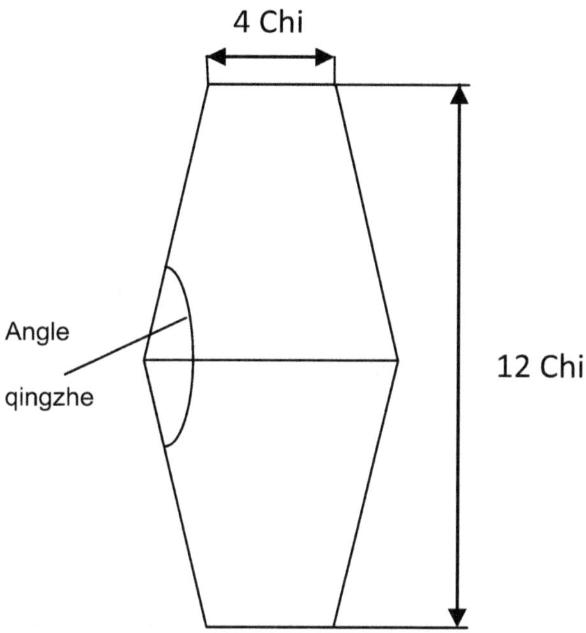

FIGURE 63 Geometry of the drum *gao* 皋
Source: Wenren Jun 闻人军, *Translated and Commented Kao Gong Ji* 考工记译注, Shanghai, Shanghai Ancient Books Press, 2008, p. 66.

Drums

The drums consisted of wooden structures that were covered with leather. A drum was shaped like a double cone, where the obtuse angle of the drum's body is 1 *qingzhe* 磬折 (151.875°). We have just encountered this angle when discussing the shape of chime stones. The definition of this angle is explained in the commentary on metrology. The geometry of the drum *gao* 皋, mentioned in "Kao Gong Ji", is presented in Fig. 63.

It is likely that, based on empiric knowledge, it was recognised that a drum with a large diameter that was small in height delivered a high, piercing tone, whereas the tone of a drum with a small diameter that was large in height was deep.

Wenren Jun has analysed the mechanism of this phenomenon mentioned in "Kao Gong Ji".[19] The drum is covered in leather on both ends, and the

19 Wenren Jun 闻人军, *"Kao Gong Ji" zhong shengxue zhishi de shuli quanshi,* 《考工记》中声学知识的数理 诠释 (*Explanation of the phonetics knowledge in "Kao Gong Ji"*), Hangzhou daxue xuebao, Ziran kexue ban, 1982, vol. 4.

inside of the drum represents a column of air. When they come into contact with this column of air, both ends of the drum vibrate and emit a tone. In thick and short drums, the air column is short, the contact between the drum's sides and the air is intense, and the dampening and loss of sound are considerable. Therefore, the vibration of the drum fades away quickly. The shorter the drum, the larger the number of acoustic wave reflections within the drum; the higher the frequency of the acoustic wave, the easier it will be absorbed when it reaches the air. The dampening of sound then happens quite fast, and the distance at which it is spread is relatively short. To the contrary, a slim and long drum has a long air column, and the contact between its sides and the air is less intense. Thus, the frequency of the drum's tone is low, the tone spreads easily, so that it is absorbed only a little by the air. The dampening is quite slow, and the distance that the tone travels will be somewhat longer.

7 Jade

Ritual objects and jewellery from jade gained special significance in Chinese art. Jade is not a mineral in the narrow sense, but a generic term for jadeite, nephrite or serpentine. The colours of jade are varied; green jade was most often used, but yellow, black-green spotted, grey and white jade were held in high esteem as well. Jade's hardness ranges from 6 to 6.5 on the Mohs scale. The material is tough and difficult to process.

As early as during the Neolithic, status objects from jade were manufactured in China. They were found in the northern Chinese Hongshan culture 红山文化 (ca. 4000–3000 BCE) and in the Liangzhu culture 良渚文化 (ca. 3300–2200 BCE) situated around the lower course of the Yangtse river.

With the development of ancestral worship, life-preserving forces were attributed to jade. Therefore, tombs of the rulers often contain many jade burial offerings. In the tomb of the female ruler Fuhao 妇好 of the Yin 殷 dynasty, which was discovered near Anyang 安阳, 755 different jade objects were found.[20]

Xia Nai 夏鼐 divided jade objects into ritual jades, weapons, tools and jewellery. The ritual jades are traditionally classified into six main shapes, which are presented in Fig. 64.

20 Xia Nai 夏鼐, *Shangdai yuqi de fenlei, dingming he yongtu* 商代玉器的分类，定名和用途 (Classification, nomenclature and use of jade objects of the Shang dynasty), Kaogu 考古, 1983, 5, pp. 455–467.

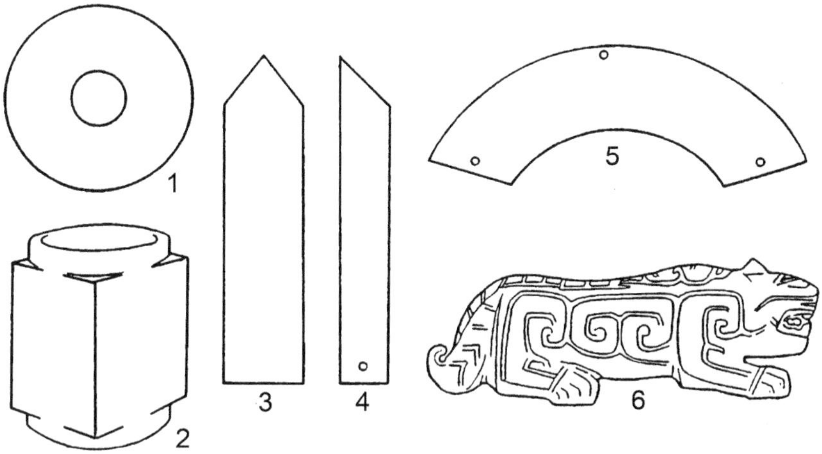

FIGURE 64 Six types of ritual jades used by the princes (jades *rui* 瑞)
Source: Xia Nai, ibid.

1 Jade disc *bi* 璧
2 Jade tube *cong* 琮
3 Jade sceptre *gui* 圭
4 Jade sceptre *zhang* 璋
5 Circular ring segment from jade *huang* 璜
6 Jade animal figure *hu* 琥

The six types of ritual jades are described in "Kao Gong Ji". The work deals with the different sizes of these jade objects, which were dependent on the status of the owner, and their general use.

Ritual jades embody the gods, so that with the jade prayers could be directed to the gods. Moreover, the jade embodies the virtue of the ruler due to its homogeneity, stability, hardness, smoothness and warmth.

It is noteworthy that several jade objects were also used as measurement standards because of their stability. It is written that the earth sceptre *tu gui* 土圭, with a length of 1 Chi 尺, 5 Cun 寸 (29.5 cm), serves to measure the shadow length. At first glance, this value seems to be too small for this aim. But in the book "Diguan" 地官, chapter *Dasitu* 大司徒 of the book *Rites of Zhou* ("Zhou Li" 周礼), it is stated: "The location where the shadow length of a gnomon of 8 Chi 尺 high (ca. 158 cm) at solstice amounts to 1 Chi 尺, 5 Cun 寸 is the centre of the earth". If the shadow length of a gnomon with 8 Chi 尺 height was thus equal the length of the jade sceptre *tu gui* 土圭, the conclusion was drawn that one was in the centre of the earth. Shadow rods with a length of 8 Chi 尺 are passed on from a later period; these were manufactured from bronze.

FIGURE 65
Jade disc *bi* 璧 (late Shang 商 dynasty (13.–10. century BCE)) Source: *Das Alte China – Menschen und Götter im Reich der Mitte* (Ancient China – Men and gods in the Middle Kingdom), Hirmer München, 1995, p. 278.

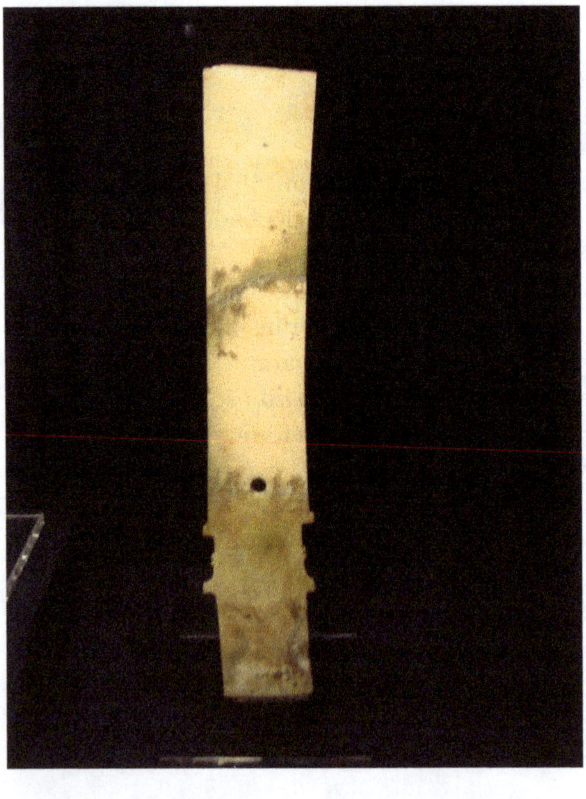

FIGURE 66
Jade sceptre *gui* 圭 (late Shang 商 dynasty, 13th–11th-century BCE) Shanghai Museum

FIGURE 67
Jade tube *cong* 琮 (Liangzhu culture 良渚文化, ca. 3000–2800 BCE)
Source: Ibid., p. 199.

FIGURE 68
Jade jewellery *huang* 璜 with two dragons (Zhanguo 战国 period)
Shanghai Museum

FIGURE 69
Jade tiger *hu* 琥 (late Shang 商 dynasty, ca. 1300–1030 BCE)
Source: *Das Alte China – Menschen und Götter im Reich der Mitte* (Ancient China – Men and gods in the Middle Kingdom), Hirmer München, 1995, p. 245.

FIGURE 70
Measurement of the shadow length with a gnomon (Qing 清 dynasty illustration from the book Shujing 书经)
Source: J. Needham, *Science and Civilisation in China*, Cambridge at the University Press, 1959, Vol. III, p. 285.

In Fig. 70, the rod lying on the ground is a jade sceptre *tu gui* 土圭.

The background to the statement that the shadow length at the centre of the earth amounts to 1 Chi 尺, 5 Cun 寸 is the following. In ancient China, the earth was imagined as a flat square that extended of 30,000 Li 里 (ca. 16,770 km) in each direction. It was assumed that the shadow length changed by 1 Cun 寸 for every 1000 Li 里 (ca. 560 km). If in the area furthest to the south, the shadow length is zero, then in the most northern part of the earth, the shadow length is 3 Chi 尺. Consequently, in the centre of the earth, the shadow length should be 1 Chi 尺, 5 Cun 寸.[21] As early as during the Sui 隋 dynasty, Liu Zhuo 刘焯 expressed his doubt in the year 604 as to the value for the change of the shadow

21 Guan Zengjian 关增建, *Zhongguo tianwenxue shi shang de dizhong gainian* 中国天文学史上的地中概念 (The concept of the centre of the earth in Chinese astronomical history), Ziran kexue shi yanjiu 自然科学史研究, vol. 19, 2000, 3, pp. 251–263.

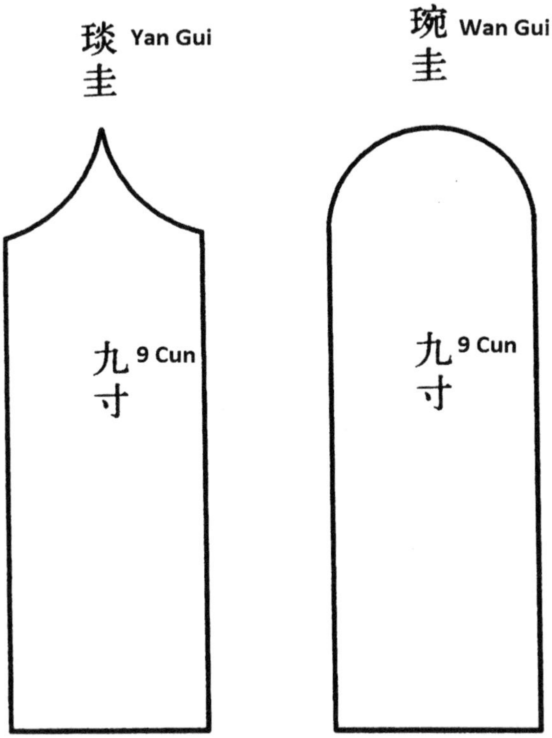

FIGURE 71
Reconstruction of the sceptres Wan Gui 琬圭 and Yan Gui 琰圭 by Dai Zhen in his work "Kao Gong Ji Tu" (*Illustrated Kao Gong Ji*)

length of 1 Cun 寸 per 1000 Li 里 being correct. He therefore proposed to organise a geodetic measurement. It was only in the following Tang 唐 dynasty that his idea was realised, through a geodetic measurement conducted by Yi Xing 一行 and Nangong Yue 南宮說 in the years between 723 and 726. This measurement took place over a length of ca. 3,500 km, reaching from Tielehuihebu 铁勒回纥部 (near Mongolia's capital Ulan Bator) to Linyi 林邑 (near the city Hue in Vietnam).[22] After the arrival of the Jesuits to China, who brought with them the knowledge that the earth is a sphere, the concept of the flat earth gradually disappeared.

It is interesting that some jade objects obviously existed only in the imagination of the author of "Kao Gong Ji". These include the sceptres Wan Gui (琬圭) and Yan Gui (琰圭). The scholar Dai Zhen (1724–1777) has provided an idea of the shape of these sceptres in his work "Kao Gong Ji Tu" (*Illustrated Kao Gong Ji*) (see Fig. 71).

22 www.Baidu.com, article Yi Xing 一行, searched on 22 March, 2018.

8 Ceramics and Porcelain

The invention of ceramics was obviously prompted by the fact that pieces of clay that were put in a fire became extraordinarily tight and hard. People have made use of this property since the Neolithic age, in order to make vessels from clay. The oldest ceramic vessels excavated in China are over 8000 years old.

The difference between ceramics and porcelain lies in the sintering behaviour of the raw materials and the firing temperature. Ceramics were made from clay earth, a weathering product of feldspar. Whereas the feldspar share in ceramics crystallises and is fired at a temperature from 800 °C to 1000 °C, porcelain consists of a mixture of kaolin, feldspar and quartz, in which the feldspar share at a firing temperature of at least 1200 °C does not crystallise again, and therefore represents an undercooled melt. If a piece of porcelain is struck, it gives a high, clear tone, whereas ceramics sound dull.

As early as during the Shang 商 and Zhou 周 dynasties, proto-porcelain was manufactured at a firing temperature set between 800 °C and 1200 °C. Ceramics workshops and kilns were discovered dating back to this time.

The "Kao Gong Ji" is the oldest written testimony in which different types of ceramic vessels are described. In particular, the book mentions the vessel types *yan* 甗, *pen* 盆, *zeng* 甑, *li* 鬲, *gui* 簋 and *dou* 豆. The vessel *yan* 甗 consists of a water container standing on three feet, with a steaming vessel located above it with holes or slits in the bottom. This is where the food that was to be steamed was placed.

The steaming jar *zeng* 甑 resembles the vessel *yan* 甗 in terms of its structure. The steamer insert had either holes or slits.

The vessel *li* 鬲 served to cook liquid food. Its three feet made it possible to put it over a fire.

The ceramic cooking jar type *li* 鬲 was used to develop the food sacrifice vessel type *ding* 鼎, made from bronze.

The vessel *gui* 簋 represents a tureen with a lid, and was used to store food or put it on the dining table.

The vessel *dou* 豆 is most often a tureen with a lid standing on a high foot. It also served to store food. The rim at the top of the vessel in Fig. 76 served to hold the lid.

The record on drinking vessels made by the carpenter Ziren 梓人 contains the statement that 4 Sheng 升 equal 1 Dou 豆. This is remarkable insofar that in the chronicle *Zuozhuan* 左传, *Third year of Duke Zhao*, this relationship between the volume units is mentioned as being part of the old system: it was the system of volume units proclaimed by the reigning duke of Qi. Later, when the Tian clan usurped power, they introduced a relationship of 10 Sheng = 1 Dou.

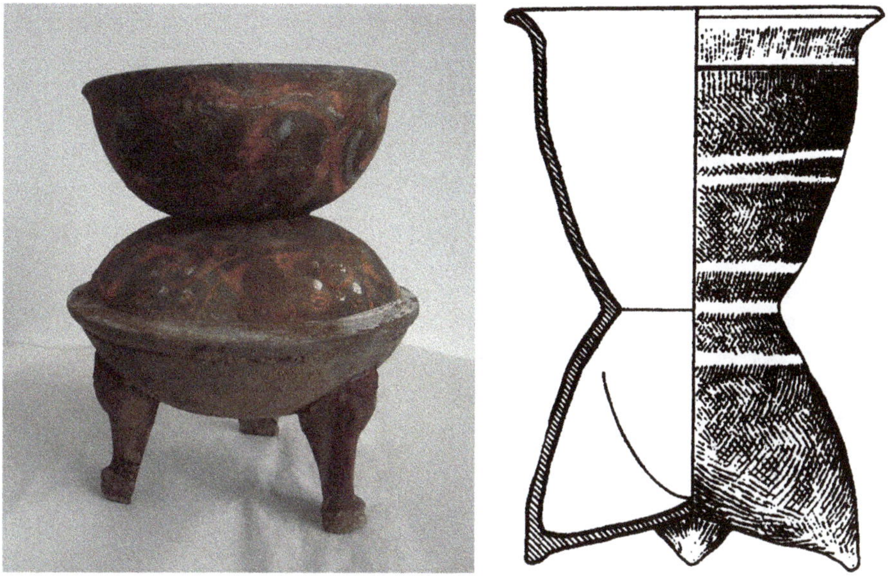

FIGURE 72 Cooking jar type *yan* 甗 (Zhanguo 战国 period) and cross section through such a jar
Source: www.Baidu.com, article yan 甗, searched on 23 March, 2018, and Dai Wusan 戴吾三, *Kao Gong Ji tushuo* 考 工记图说 (Illustrated and Explained Kao Gong Ji), Shandong huabao chubanshe 山东画报出版社, 2003, p. 68.

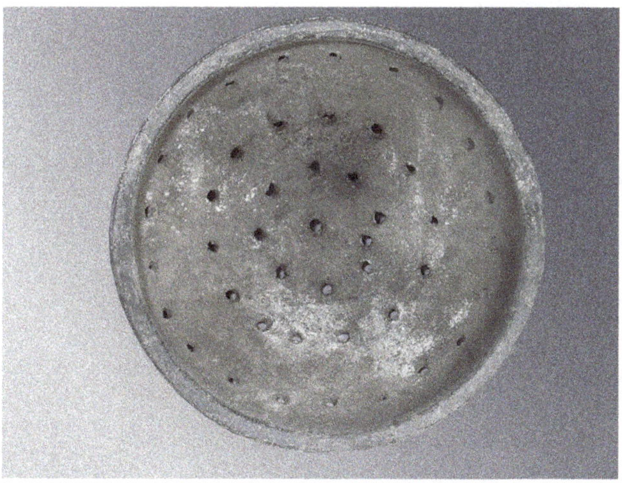

FIGURE 73 Steamer insert of cooking jar type *zeng* 甑 (Zhanguo 战国 period)
Source: www.Baidu.com, article zeng 甑, searched on 23 March, 2018.

FIGURE. 74
Cooking jar type *li* 鬲, excavated at Erligang 二里岗 (Neolithic age)
Source: *Zhongguo wenwu jinghua da cidian* 中国文物精华大辞典 (Encyclopaedia of the essence of Chinese cultural assets), vol. Ceramics and Porcelain, Shanghai cishu chubanshe, 1995, p. 60.

FIGURE 75 Cooking jar of type *gui* 簋 (Shang 商 dynasty)
Source: Encyclopaedia of the Essence of Chinese Cultural Assets, Shanghai: Shanghai Lexicigraphic Publishing House, Hongkong: The Commercial Press, Joint Publishing, 1985.

FIGURE 76
Ceramic vessel type *dou* 豆 (Zhanguo 战国 period)
Source: www.Baidu.com, article dou 豆, searched on 23 March, 2018.

FIGURE 77 Ceramic vessel of type *pen* 盆 (Yangshao culture 仰韶文化, 5000–3000 BCE, excavated at Banpo 半坡 village near Xi'an 西安 1954)

Also, a system of quality control is described: a supervisor punished the craftsperson in case the quality standards were not fulfilled.

The "Kao Gong Ji" mentions two officials that were responsible for the manufacture of ceramics: *Taoren* 陶人 and *Fangren* 旊人. Scientists interpret the division of labour between these two as follows: *Taoren* 陶人 was responsible for everyday ceramics and *Fangren* 旊人 for ritual vessels.[23] Moreover, they state that *Taoren* 陶人 used a potter's wheel and *Fangren* 旊人 a form gauge.[24] Finally, the difference between them also lay in the fact that *Taoren* 陶人 manufactured ceramics and *Fangren* 旊人 proto-porcelain.[25] Future archaeological discoveries will possibly provide us with more clarity about the division of labour between these two officials.

It is remarkable that the "Kao Gong Ji" mentions a gauge for the profile of the vessels under the supervision of the official *Fangren* 旊人. This gauge clearly served to guarantee the consistency of the shape and volume of the vessels.

9 Civil Engineering

The "Kao Gong Ji" describes the following fields of civil engineering:

- Layout of cities and palaces
- Construction of houses
- Construction of irrigation channels

The architecture of ancient China is primarily characterised by timber construction, whereas in Europe, representative buildings were made from stone.

Determining Direction

When a capital or palace was planned, the main axis was oriented in a north-south direction, in order to establish the unity between heaven and earth. Therefore, methods were needed to determine direction. The "Kao Gong Ji" describes a method that entails a gnomon measuring the shadow length at sunrise and sunset. The intersection points of the shadows are determined through a circle around the gnomon. The connecting line of both intersection points then provides the east-west direction (see Fig. 78). The shadow at noon

23 Kong Yingda 孔颖达, *Commentary to Liji Qulixia* 礼记曲礼下, in: Shisanjing zhushu 十三经注疏 (The 13 classics with explanations and commentaries), Zhonghua shuju 中华书局, 1980, p. 1261.

24 Zhu Yan 朱琰: *Tao shuo* 陶说 (Explanations on ceramics), Translation and commentary by Fu Zhenlun 傅振伦, Qinggongye chubanshe 轻工业出版社, 1984, pp. 8–59.

25 Wang Qingzheng 汪庆正: *Zhongguo taoci shi yanjiu zhong ruogan wenti de tansuo* 中国陶瓷史研究中若干问题的探索 (Investigation of several problems in the research of the history of porcelain in China), Shanghai guji chubanshe 上海古籍出版社, 1983.

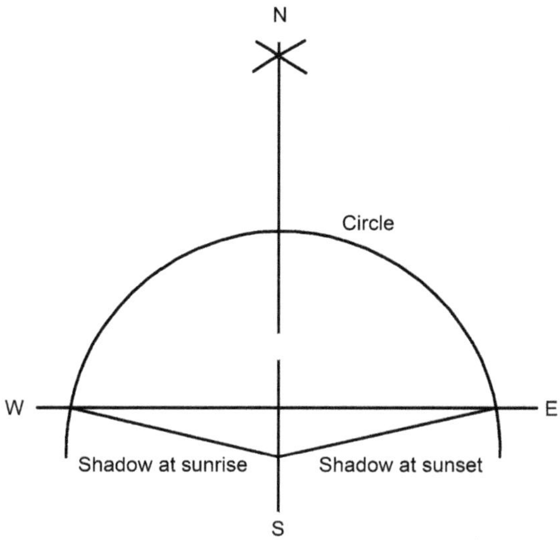

FIGURE 78
Method described in "Kao Gong Ji" for the determination of the east-west direction
Source: Dai Wusan 戴吾三, *Kao Gong Ji tushuo* 考工记图说 (Illustrated and Explained Kao Gong Ji), Shandong huabao chubanshe 山东画报出版社, 2003, p. 79.

and the Pole Star at night also provide references used to determine the south-north direction. In this way, the four directions could be determined.

Layout of Cities

The "Kao Gong Ji" lays down rules for the layout of the:
- king's city
- cities of the dukes
- cities of the dignitaries

These rules reflect the societal hierarchies during the early Zhou 周 dynasty.

The king's city distinguishes itself by the following settings:
- The city measures 9 Li 里 square, with 3 city gates on each side.
- The city is crossed by 9 main streets in a N–S and 9 in an E–W direction; the main streets have a width of 9 carriage track widths gui 轨; the street along the city wall has a width of 7 carriage tracks; and the streets of the suburbs a width of 5 carriage tracks.
- The temple of the ancestors is on the left, and the temple of the earth and the millet god is on the right of the symmetry axis of the palace. The palace is in front, and the market in the back.
- The inner palace consists of 9 buildings, in which 9 concubines live. The outer palace consists of 9 buildings, in which the 9 ministers conduct the affairs of state.

The "Kao Gong Ji" prescribes the following rules for the king's palace:
- Palace and market have an area of 100 bu 步 square each.
- The wall around the palace is 5 Zhi 雉 (Zhang 丈) high; the corner towers of the palace wall are 7 Zhi 雉 high; and the palace gates are 5 Zhi 雉 high.

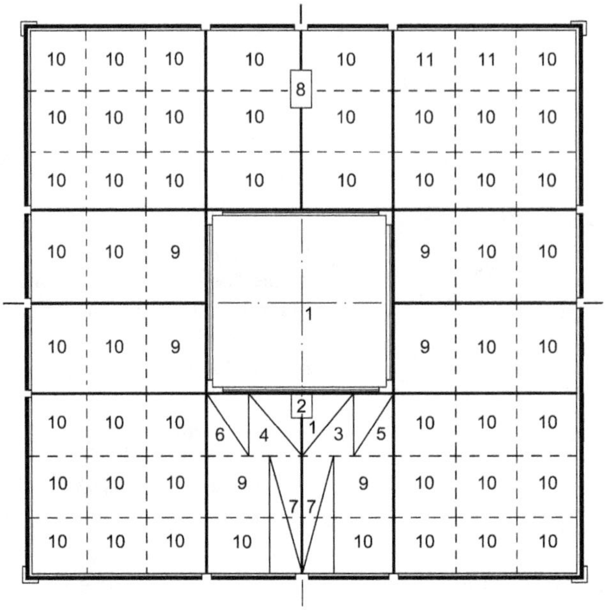

1 Palace
2 Forecourt
3 Temple of ancestors
4 Temple of earth and millet god
5 King's warehouse
6 Horse stables
7 Houses of ministers
8 Market
9 Houses of officials
10 Living quarters
11 Granary

City wall
Main street
Side street
Ring street

FIGURE 79 Reconstruction of the layout of a king's city during the Zhou 周 dynasty
Source: *Dai Wusan, Illustrated and Commented Kao Gong Ji*, Shandong Huabuao Chubanshe, 2003, p. 124

These statements are partly confirmed and partly complemented by writings from the Chunqiu 春秋 and the Zhanguo 战国 periods, for instance the *Book of Documents* ("Shujing" 书经), the *Book of Songs* ("Shijing" 诗经) and the *Discourses on the States* ("Guoyu" 国语), as well as by inscriptions on bronze vessels from this time.

Fig. 79 represents an attempt to reconstruct a king's city according to the statements of the "Kao Gong Ji" and other mentioned sources, while Fig. 80 is an attempt to reconstruct a king's palace from the Zhou 周 dynasty.[26]

26 He Yeju 贺业钜: *"Kao Gong Ji" yingguo zhidu yanjiu* 考工记营国制度研究 (Investigations of the arrangement in city building according to "Kao Gong Ji"), Zhongguo jianzhu gongye chubanshe 中国建筑工业出版社, 1985, p. 51, p. 97.

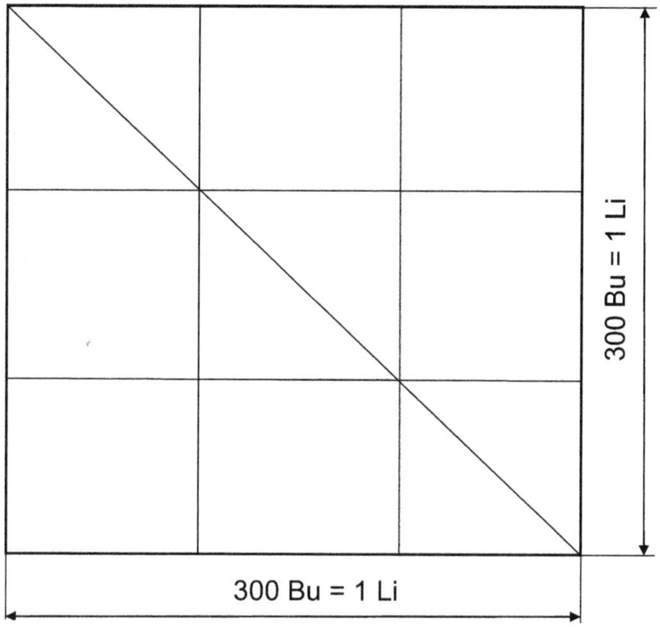

FIGURE 80 Area module of the city layout derived from the well-field system

The area modules of the city layout derived from the well-field system consist of squares of 1 Li 里 by 1 Li (resp. 300 Bu 步 by 300 Bu), as shown in the following scheme.

An interesting detail is found in the layout of the king's ancestor temple. The somewhat cryptic description in "Kao Gong Ji" can be interpreted as follows.[27] The temple consists of five rooms, which symbolise the five elements (*wu xing* 五行). In the middle is the room dedicated to the element Earth (*tu*) 土 and on the four corners we find, in the north-east, the "wood" (*mu*) 木 hall; in the south-east, the "fire" (*huo*) 火 hall; in the south-west, the "metal" (*jin*) 金 hall; and in the north-west, the "water" (*shui*) 水 hall.

The plans for the cities of the dukes and dignitaries are derived from the rules for the king's city in a hierarchical way, so that they embody their subordination to the king. Correspondingly, in the city of a vassal duke, the corner towers of the city wall are only allowed to be 7 Zhi 雉 high, and in the city of the dignitaries, only 5 Zhi, as opposed to 9 Zhi in the king's city.

The main streets in the city of the vassal dukes are only allowed to be 7 carriage tracks gui 轨 wide, and in the city of the dignitaries only 5 carriage

27 Guan Zengjian, *Kao Gong Ji jiang ren ying guo* 考工记匠人营国 (Civil engineering in Kao Gong Ji, personal communication).

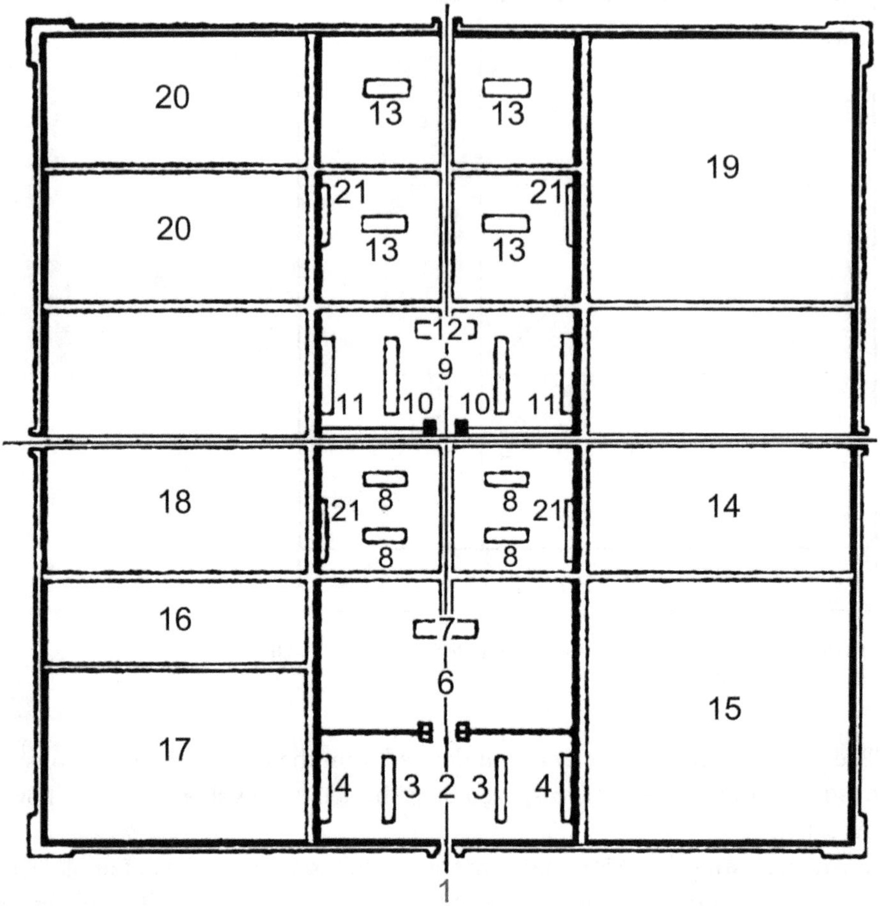

1 palace gate
2 court of audiences
3 houses of nine dignitaries
4 houses of officials
5 gate to the inner chambers
6 swallow court (court of concubines)
7 king's chambers
8 king's swallow chambers
9 court of northern palace
10 houses of the nine concubines

11 houses of the female servants
12 official rear chambers
13 small rear chambers
14 palace of crown prince
15 palace of the king
16 area of officials
17 granary
18 kitchen
19 area of female servants
20 area of craftsmen
21 clothing store

FIGURE 81 Reconstruction of the layout of a king's palace during the Zhou 周 dynasty
Source: *Dai Wusan, Illustrated and Commented Kao Gong Ji*, Shandong Huabao Chubanshe, 2003, p. 125

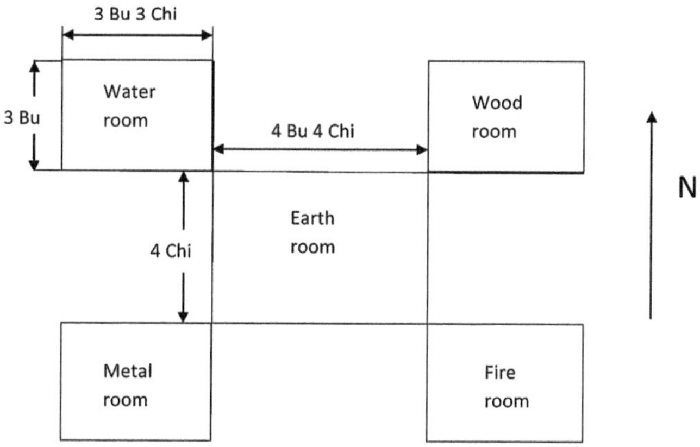

FIGURE 82 Layout of the king's ancestor temple

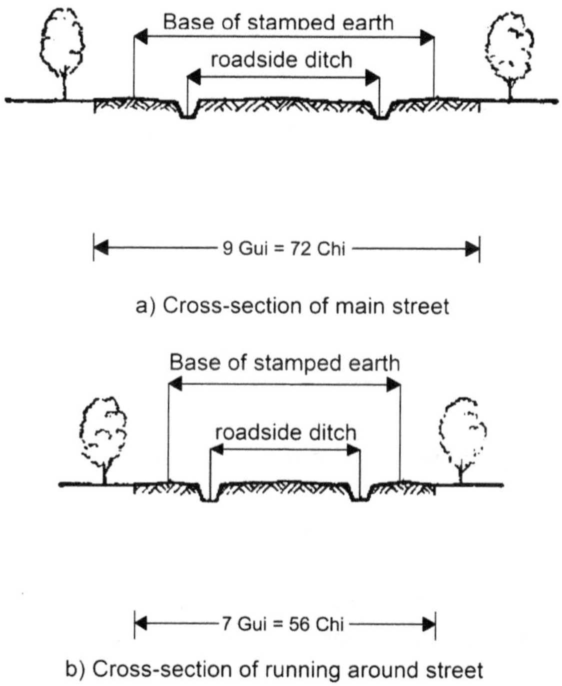

FIGURE 83 Cross section of main streets (above) and of the street running along the city wall of the king's city
Source: Dai Wusan 戴吾三, *Kao Gong Ji tushuo* 考工记图说 (Illustrated and Explained Kao Gong Ji), Shandong huabao chubanshe 山东画报出版社, 2003, p. 126.

tracks, as opposed to 9 carriage tracks in the king's city. One carriage track corresponds to 8 Chi = 157 cm.

Fig. 83 shows a cross section of the main streets and the street running along the city wall of the king's city.

The chessboard-shaped layout of the city is probably derived from the well-field system of agriculture that was in use during the Zhou 周 dynasty. Based on the character *jing* 井, which means "well", the fields were divided into 3-by-3 sections, where the yield from the field in the middle had to be handed over to the ruler.

Construction of Houses

The "Kao Gong Ji" essentially deals with the fact that thatched roofs should be steeper than tiled roofs. For thatched roofs, the height of the roof should be one third of the house's depth. On the other hand, on tiled roofs, this ratio should amount to one quarter. In this way, it is guaranteed that the rainwater runs off without soaking the roof.

Construction of Irrigation Channels

The system of irrigation ditches is based on the rule that the width and depth of the ditches should double at every 30 Li (ca. 14 km) each. Correspondingly, we obtain the scheme provided in the following table.

TABLE 5 Kinds of irrigation ditches

Designation of irrigation ditch	Width and depth
Rut (*quan* 甽)	1 Chi = 20 cm
Groove (*sui* 遂)	2 Chi = 39 cm
Gutter (*gou* 沟)	4 Chi = 79 cm
Drain (*xu* 洫)	8 Chi = 1.57 m
Channel (*kuai* 浍)	16 Chi = 3.14 m

Obviously, we have to do with approximate values here, which must be adapted to the topography.

The barrage in a watercourse described in "Kao Gong Ji" could have the following design (Fig. 84).

It is interesting that in the shape of the barrage the angle *qingzhe* 磬折 appears again, which is used like a standardised obtuse angle.

Referring to the realisation of the rules for the layout of cities as provided in the "Kao Gong Ji", so far no cities from the Chunqiu 春秋 and the Zhanguo 战国 periods that corresponded to these rules were excavated. Cities from the Zhanguo 战国 period, such as the excavated capital Linzi 临淄 of the principality Qi 齐, distinguish themselves above all by the fact that they provide much more space for the activity of the craftsmen and merchants

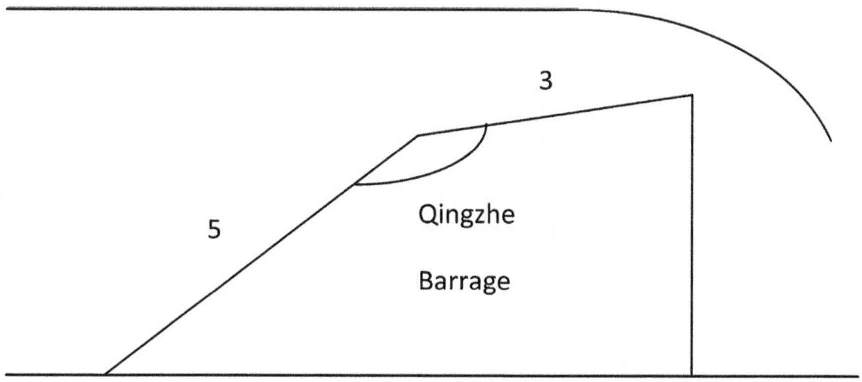

FIGURE 84 Design of a barrage in a watercourse
Source: Wenren Jun 闻人军, *Kao Gong Ji yizhu* 考工记译注 (Translation and annotation of Kao Gong Ji), Shanghai guji chubanshe, 2008, p. 122.

compared with the description in the "Kao Gong Ji". This reflects the significant economic development of this time, and the fact that crafts and trade no longer prioritized the desires of the ruler and the dukes, but of urban society as a whole. Moreover, the "Kao Gong Ji" was lost during this time, and was only rediscovered during the Han 汉 dynasty. Therefore, it is not surprising that the remnants of the capital Chang'an 长安 from the Han 汉 dynasty embody a reflection of the spirit of "Kao Gong Ji". Because of the terrain, the city is certainly not planned as a square, but features such as the orientation along a north-south axis, three city gates on each side and streets running perpendicularly correspond to the demands of "Kao Gong Ji" (see Fig. 85).

The following figure shows the reconstruction of the capital Chang'an 长安 during the Tang 唐 dynasty, based on a representation engraved in stone from the year 1080.

During the Tang 唐 dynasty, Chang'an 长安 was inhabited by approximately one million people. As it was the final stop of the famous silk route, many caravans of foreign merchants travelled to this city. These merchants lived in special quarters.

The square layout of the city corresponds to the description of "Kao Gong Ji", as do the location of the emperor's palace in the north of the city, the streets running in a north-south and east-west direction, and the three gates on each side of the city wall. Nevertheless, the markets are not situated north of the palace, but on the northern end of the living quarters, which was better adapted to the needs of the people.

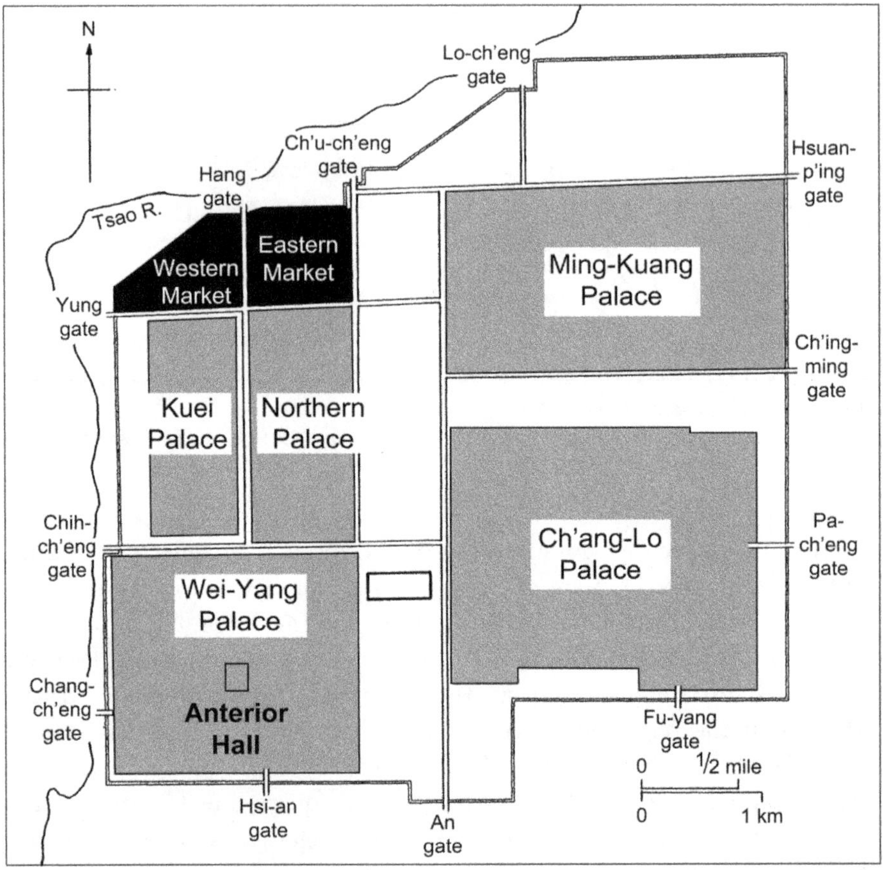

FIGURE 85 Layout of the capital Chang'an 长安 during the Han 汉 dynasty
Source: D. Twitchett, John K. Fairbank (ed.), *The Cambridge History of China*, Vol. 1: The Ch'in and Han Empires 221 BC–AD 220, Cambridge University Press, 1986, p. 134.

The influence of "Kao Gong Ji" can also be detected in the layout of the capital Beijing 北京 as it existed around the year 1900. Present-day Beijing was constructed in the years of Ming 明 emperor Yongle 永乐 in the beginning of the 15th century.

According to the description in "Kao Gong Ji", the imperial palace is designed along an axis running in a north-south direction. This axis is also found in the layout of the remaining city. The Manzu 满族 people, who were privileged during the Qing 清 dynasty, were allowed to live around the imperial palace, whereas the Chinese had to live in the Chinese city situated in the south. The

FIGURE 86 Reconstruction of the layout of the capital Chang'an 长安 during the Tang 唐 dynasty
Source: D. Kuhn (ed.): *Chinas Goldenes Zeitalter* (China's Golden Age), Edition Braus, 1993, p. 37.

streets are designed running perpendicularly to each other in a north-south and east-west direction.

The arrangement of the ancestor's temple on the left of the north-south axis and of the millet god temple on the right corresponds to the prescription of the "Kao Gong Ji". Furthermore, we find three gates on the east, south and west sides of the city. However, on the north side of the Tartar city only two city gates existed, complemented by two gates on the north side of the Chinese city and the Tiananmen 天安门, the Gate of Heavenly Peace, on the south side of the imperial palace.

A detail from a contemporary city map of Beijing 北京 shows that, despite the rapid development of the city, the essential features of its historical layout have been preserved.

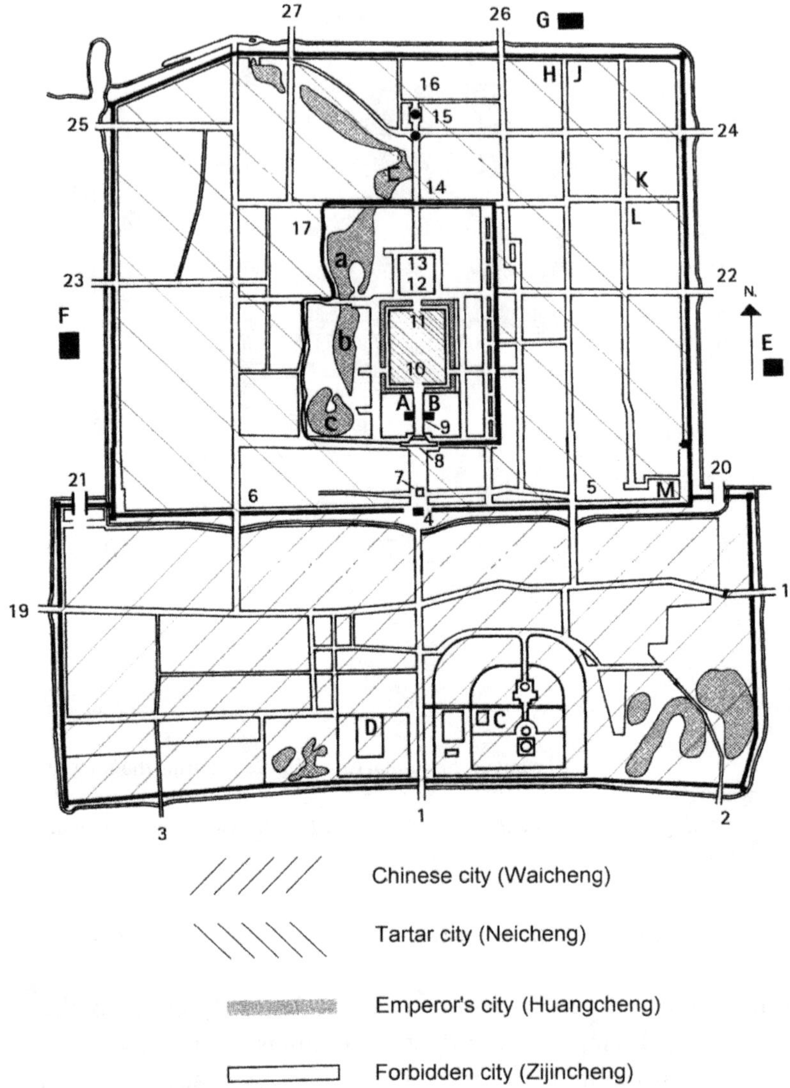

A	Imperial ancestor temple	G	Altar of the earth
B	Earth and millet god temple	H	Imperial academy
C	Temple of heaven	I	Confucius temple
D	Altar of agriculture	J	Lama temple
E	Altar of the sun	K	Imperial granary
F	Altar of the moon	L	Imperial warehouses
		M	Observatory

FIGURE 87 Layout of the capital Beijing 北京 around the year 1900
Source: Reiss-Museum Mannheim: Die Verbotene Stadt (The Forbidden City), Verlag Philipp von Zabern Mainz (1997), p. 69.

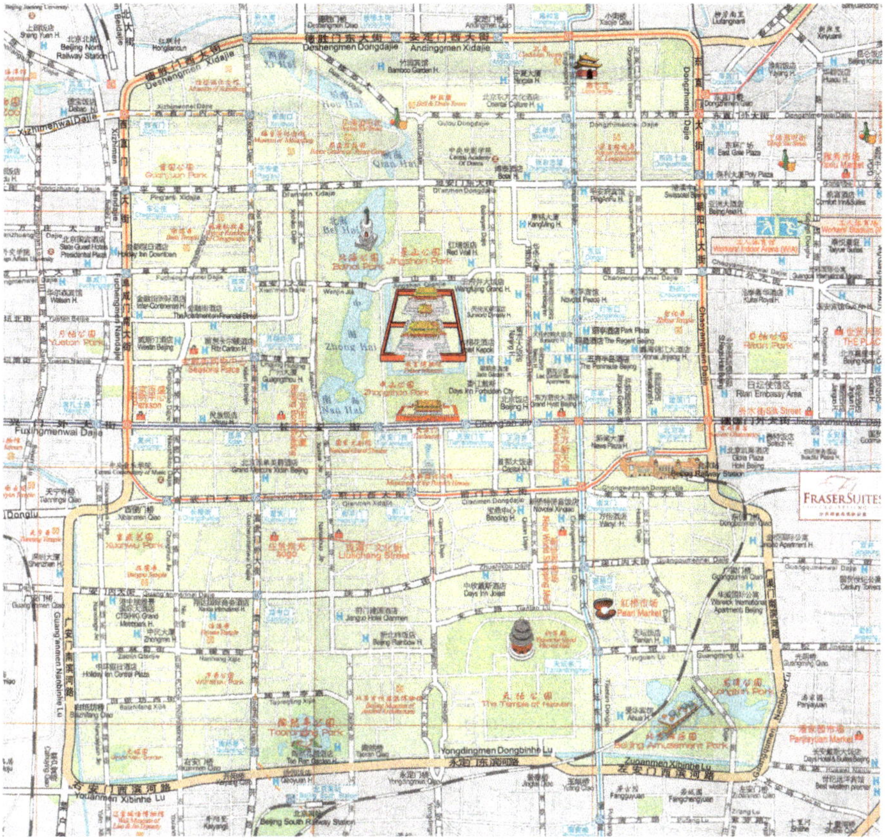

FIGURE 88 Detail of the centre of Beijing 北京 from a contemporary city map

10 Manufacture of Bows

The "Kao Gong Ji" turns its attention to the manufacture of bows and discusses the selection of materials, the technology involved, and the appropriate processing time.

Altogether, six different materials are needed:

- Wood for the bow's body
- Horn to strengthen the bow's body
- Tendons for the bowstring
- Silk to wind around the bow's body, which is strengthened with horn parts
- Lacquer as protection of the bow against humidity

For the bow's body, seven kinds of wood are listed, in the order of their suitability:

- Cudrania as the most suitable wood
- Grey oak
- Wild mulberry tree
- Mandarin tree
- Flowering quince
- Blackthorn
- Bamboo

The wood of the bow's body should be cut in winter, because during this time, the grain of the wood is regular and dense. It is remarkable that for all materials in "Kao Gong Ji", the correct time of processing is indicated, which is at least partly justified. On no account can the grain be hurt. Possible knots in the wood have to be removed cautiously, because they can damage the bowstring or the horn parts due to friction.

The bow's body is soaked in water twice, so that it can be bent easily. Finally, it is bent over a flame, where the flame should not be too hot.

For the horn parts, the horns of healthy, not too old cattle should be used, and the cattle should be slaughtered in autumn. The horns should be as straight as possible, because long, straight horn parts are needed to strengthen the bow's body. In addition, the tip of the horn should be as blunt as possible, because it is most flexible in that way. The horn is soaked three times in spring, so that it is as flexible as possible for processing.

For the bowstring, narrow, long strips and larger, round pieces are needed. The latter obviously serve to fix the bowstring to the bow's body. The tendons should be processed in summer, because then they will not stick together.

The glue is obtained from the following animals:

- Deer
- Horse
- Cattle
- Rat
- Fish

For each kind of animal glue, the standard colour is indicated. In autumn, the bow's body should be joined together with the horn parts, silk should be wound around it, and the glue should be applied evenly.

Regarding the lacquer, it is only written that it should be clear.

The silk is required to shine.

The schematic structure of a bow is displayed in the following picture.

Besides the requirements for strength, durability and freedom of warp of the bow described above, the force of the bow is an important criterion as well.

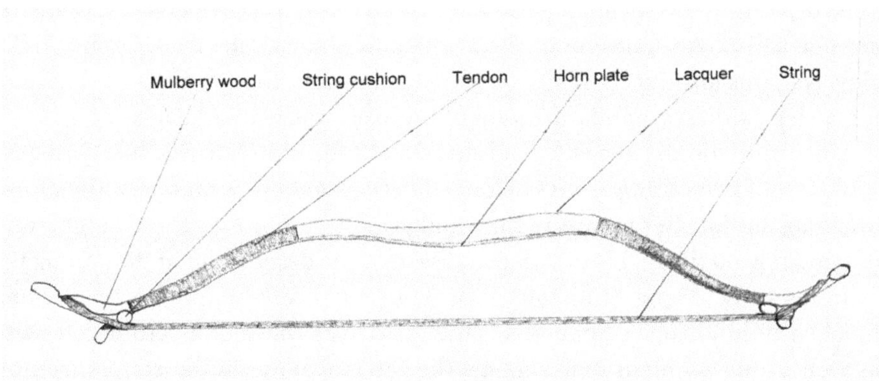

FIGURE 89 Structure of a bow
Source: Wenren Jun 闻人军, *Kao Gong Ji* 考工记, Zhongguo guoji guangbo chubanshe 中国国际广播出版社, Beijing 北京, 2010, p. 275.

On this topic, it is written that a weight of three Jun 钧 (18 kg) is hung on the bowstring. Obviously then, the deflection of the bowstring is measured, but this is not described.

According to the social status of the owner, four different large bows are distinguished:

TABLE 6 Lengths of bows according to social status of the owner

Owner of the bow	Length of the bow
Son of heaven (king)	1/9 of the circumference
Vassal duke	1/7 of the circumference and 6 Chi 尺, 6 Cun 寸 (130.0 cm)
Dignitary	1/5 of the circumference and 6 Chi 尺, 3 Cun 寸 (124.1 cm)
Official	1/3 of the circumference and 6 Chi 尺 (118.2 cm)

It is interesting that length is indicated as a part of a circumference (but the diameter of the circle is unknown). On the other hand, length can be derived from the specified chord length of part of the circumference. If length is given as a part of the circumference, then the idea of the full circle as the reference quantity for an angle measure is applied. But on the basis of the full circle, no angle measurement unit was defined.

In the chapter in "Kao Gong Ji" on bronze weapons, the manufacture of bronze arrowheads is discussed. It is stated that the arrowheads are one Cun 寸

(2.0 cm) long and have a circumference of one Cun. The wooden arrow shafts are one Chi 尺 (19.7 cm) long. The best arrow shafts come from Fenhu 妢胡. However, this locality could not be identified clearly. Due to the large demand for such products, a standardization of the sizes was realised early.

Since until the last dynasty that ruled over China, the Qing 清 dynasty (1644–1911), bows for shooting arrows belonged to the standard weapons of the army, one finds in the later literature time and again that hints on the manufacture of bows rely on the highly detailed and systematic description in "Kao Gong Ji".

In his work "Meng Xi Bi Tan" 梦溪笔谈 (*Brush talks on the dream stream*, finished around 1090), Shen Kuo 沈括 discusses the manufacture of bows, as well as the problem that data on the force of bows in old scripts cannot be compared directly due to the change of the values of measurement units, and would have to be converted. Shen Kuo listed six properties that a bow should have:[28]

- Bows should be light and firm
- Bows should be drawn easily
- The shooting force should be preserved for a long time
- The tone of the bowstring should sound clearly
- The drawn bow should not twist

As pointed out in "Kao Gong Ji", Shen Kuo mentions that the bow's body should be reinforced with cattle horn in the grip area, with a tendon wound around it and finally glued together.

In order for the bow not to twist, the grain of the wood must be straight.

The technological encyclopaedia "Tian Gong Kai Wu" 天工开物 (*The Exploitation of the Works of Nature*) by Song Yingxing 宋应星 from the year 1637 contains a detailed description of the manufacture of arrows and bows.[29] It deviates from "Kao Gong Ji" in that bamboo is preferred for the bow's body. Here, one must consider that the "Kao Gong Ji" deals with products for the needs of the ruling class, especially the nobles. Therefore, it is quite conceivable that the bows for the king and the vassal dukes were made from the relatively rare cudrania wood, and the bows for the soldiers from the cheaper, more easily available bamboo.

The "Tian Gong Kai Wu" indicates that the bamboo is cut in autumn, because bamboo cut in spring and summer is often infested with insects. The bow's body is also reinforced with cattle horn, with cattle tendons wound

28 Hu Daojing 胡道静 (Ed.), Shen Kuo 沈括: *Meng Xi Bi Tan jiaozheng* 梦溪笔谈校正 (Commented edition of "Brush talks on the dream stream"), Bashu shushe 巴蜀书社, 1988.

29 Zhong Guangyan 钟广言 (Ed.): *Tian Gong Kai Wu* 天工开物 (The Exploitation of the Works of Nature), Guangdong renmin chubanshe 广东人民出版社, 1976.

around it, and stuck together with fish glue. An innovation is that on the grip's location of the bow's body, soft birch bark is glued.

The bowstring is no longer manufactured from cattle tendons, but from silk. The silk threads are wound around a core of more than 20 silk threads. They are obtained from silk worms fed with the leaves of cudrania trees, since these are firmer and tougher than mulberry silk trees. Compared with strings made from cattle tendons, silk bowstrings have the advantage that they do not loosen in highly humid air. The "Tian Gong Kai Wu" mentions that before the northern neighbouring people had copied this innovation from the Chinese, they did not dare carry out belligerent attacks during rain and fog, because they used cattle tendons for the bowstrings originally.

Great importance is attached to the slow drying of the bow's body and to cover the finished bow in glue. Also, when storing the finished bow, it is important to avoid mildew and humidity, which would decompose the glue. Therefore, in winter, the bows that were laid in a bow case were warmed up by a charcoal fire from below. Ordinary soldiers, who could not afford such expenses, hung the bow on the wall over the kitchen stove.

Arrow shafts were made from bamboo in southern China and from willow wood in northern China. Generally, they were two Chi 尺 (62 cm) long. The arrowhead was one Cun 寸 (3.1 cm) long. In this way, the arrow shafts were twice as long as indicated in "Kao Gong Ji". The reason is clearly that their longer length better stabilised their flight. It is remarkable that for the arrowheads, the same measure as in "Kao Gong Ji" is provided. However, the value of the length unit Cun 寸 enlarged one-and-a-half times until the Ming 明 dynasty, compared with the value during the Chunqiu 春秋 period.

The "Kao Gong Ji" indicates the principle of testing the bow's force with a weight of 3 Jun 均, and the deflection of the bow string is measured.

In "Tian Gong Kai Wu", the determination of the bow's force is described in more detail. The tester steps on the bowstring and hangs the bow in the middle on the hook of a sliding weight scale. When the bow is drawn fully, the weight on the scale beam shifts, and at equilibrium, one obtains the force of the bow. An original illustration from "Tian Gong Kai Wu" shows this process. However, the illustrator did not understand the test procedure correctly. Qiu Guangming 丘光明 inserted a sketch into her work[30] that reflects the description of "Tian Gong Kai Wu" 天工开物.

30 Qiu Guangming 丘光明: *Jiliang shi* 计量史 (A history of metrology), in: Zhongguo wulixue shi daxi 中国物理学史大系 (A series of books on history of physics in China), Hunan jiaoyu chubanshe 湖南教育出版社, Changsha 长沙, 2002, p. 617.

FIGURE 90
Test of the bow force (original illustration of "Tian Gong Kai Wu" 天工开物)

FIGURE 91
Corrected illustration by Qiu Guangming 丘光明 for the test of the bow force
Source: *Qiu Guangming, A Series of Books on History of Physics in China – A History of Metrology*, Hunan Jiaoyu Chubanshe, 2002, p. 617

Bibliography

Annex: Index of Research Works on "Kao Gong Ji"

Research Papers

曹佐熙，读《考工记》，船山学报，1915, (4).

Cao Zuoxi, *Reading "Kao Gong Ji"*, Chuanshan Xuebao, 1915, (4).

陈见东，亚里士多德"四因说"与《考工记》造物原则比较，装饰，2006, (12).

Chen Jiandong, *A comparison between Aristotle's doctrine of the four elements and the creation principles in "Kao Gong Ji"*, Zhuangshi, 2006, (12).

陈久金，《考工记》中的天文知识，中国科技典籍研究 – 第一届中国科技典籍国际会议论文集，郑州：大象出版社，1998。

Chen Jiujin, *The astronomical knowledge in "Kao Gong Ji", Research on the classics of Chinese science and technology*, Proc. of the 1. International conference on the classics of Chinese science and technology, Zhengzhou, Daxiang publ. house, 1998.

陈丽萍、黄伯璋，《考工记》设计理念中的天人思想，文艺争鸣，2010, (6).

Chen Liping, Huang Bozhang, *The thought of heaven and man in the design ideas of "Kao Gong Ji"*, Wenyi Zhengming, 2010, (6).

陈荣红，《考工记》– 我国现存最早的科技档案汇编，浙江档案，2004, (5).

Chen Ronghong, *"Kao Gong Ji" – China's earliest existing scientific-technological document collection*, Zhejiang Dang'an, 2004, (5).

陈筱娇，简析《考工记》、《墨子》分列的先秦工艺差异，大众文艺，2010, (15).

Chen Xiaojiao, *Short analysis of the differences of pre-Qin technologies as described in "Kao Gong Ji" and "Mo Zi"*, Dazhong Wenyi, 2010, (15).

陈学军，浅谈《考工记》中的设计思想与现代设计，国画家，2009, (5).

Chen Xuejun, *A talk on the design ideas in "Kao Gong Ji" and the present design*, Guohuajia, 2009, (5).

陈衍，考工记辨正，国学论衡，1934, (3, 04).

Chen Yan, *Debate on Kao Gong Ji*, Guoxue Lunheng, 1934, (3, 04).

陈正俊，《考工记》成文时代探议，装饰，2002, (11).

Chen Zhengjun, *Discussion of the time of origin of "Kao Gong Ji"*, Zhuangshi, 2002, (11).

陈正俊，《考工记》设计思想中的"势"，设计艺术，2002, (2).

Chen Zhengjun, *The term "shi" in the design ideas of "Kao Gong Ji"*, Sheji Yishu, 2002, (2).

陈仲先，从《考工记》的色彩观看"五色审美观"，美与时代，2006, (12).

Chen Zhongxian, *From the colours in "Kao Gong Ji" viewing the "aesthetics of the five colours"*, Mei yu Shidai, 2006, (12).

近重真澄，东洋古代文化之化学观，科学，1919, (3).

Chikashige Masumi, *Views on chemistry in ancient oriental culture*, Kagaku, 1919, (3).

戴吾三，《考工记》“匠人建国”略议，齐鲁学刊，1994, (4).

Dai Wusan, *A short discussion of the chapter "Jiang Ren Jian Guo" in "Kao Gong Ji"*, Qilu Xuekan, 1994, (4).

戴吾三，《考工记》的篇章结构分析，中国科技典籍研究 – 第二届中国科技典籍国际会议论文集，郑州:大象出版社，2003。

Dai Wusan, *Analysis of the chapter structure of "Kao Gong Ji", Research on the classics of Chinese science and technology*, Proc. of the 2. International conference on the classics of Chinese science and technology, Zhengzhou, Daxiang publ. house, 2003.

戴吾三，《考工记》中轮之检验新探，中国科技史料，2000, (2).

Dai Wusan, *New investigation of the testing of wheels in "Kao Gong Ji"*, Zhongguo Keji Shiliao, 2000, (2).

戴吾三，论《考工记》的生产技术管理，大自然探索，1996, (1).

Dai Wusan, *On the management of production technologies in "Kao Gong Ji"*, Daziran Tansuo, 1996, (1).

戴吾三，图说《考工记》，北京日报，2004 年 3 月 29 日。

Dai Wusan, *An illustrated "Kao Gong Ji"*, Beijing Ribao, 29.3.2004.

戴吾三、邓明立，《考工记》的技术思想，自然辩证法通讯，1996, (1).

Dai Wusan, Deng Mingli, *The technological thought of "Kao Gong Ji"*, Ziran Bianzhengfa Tongxun, 1996, (1).

戴吾三、高宣，《考工记》的文化内涵，清华大学学报 (哲学社会科学版), 1997, (2); 中国科技典籍研究 – 第一届中国科技典籍国际会议论文集，郑州:大象出版社，1998。

Dai Wusan, Gao Xuan, *The cultural content of "Kao Gong Ji"*, Qinghua Daxue Xuebao (Zhexue shehui kexue ban), 1997, (2); Research on the classics of Chinese science and technology, Proc. of the 1. International conference on the classics of Chinese science and technology, Zhengzhou, Daxiang publ. house, 1998.

戴吾三、张鸷中，从《考工记》看中国古代车辆的设计，中国科技典籍研究 – 第一届中国科技典籍国际会议论文集，郑州:大象出版社，1998。

Dai Wusan, Zhang Zhizhong, *From "Kao Gong Ji" explain the design of ancient Chinese chariots, Research on the classics of Chinese science and technology*, Proc. of the 1. International conference on the classics of Chinese science and technology, Zhengzhou, Daxiang publ. house, 1998.

邓学忠、姚明万、邓红潮，《考工记》中的制车手工业标准化及对秦代的影响，南阳师范学院学报，2012, (4).

Deng Xuezhong, Yao Mingwan, Deng Hongchao, *The standardization of handicraft of chariot manufacture in "Kao Gong Ji" and its influence on Qin epoch*, Nanyang Shifan Xueyuan Xuebao, 2012, (4).

董英哲、魏建忠、牛有萍、杨丙雨，《考工记》的技术理论，西北建筑工程学院学报 (自然科学版), 1994, (2).

Dong Yingzhe, Wei Jianzhong, Niu Youping, Yang Bingyu, *The technological theory of "Kao Gong Ji"*, Xibei Jianzhu Gongcheng Xueyuan Xuebao (Ziran kexue ban), 1994, (2).

杜军虎，"方色" 论析 – 由《考工记·画缋》之 "五色" 与 "六色" 并提谈起，创意与设计，2011, (4).

Du Junhu, *Analysis of "directions and colours" – Discussion of "Five Colours" and "Six Colours" in "Kao Gong Ji·Hua Hui"*, Chuangyi yu Sheji, 2011, (4).

杜正国，《考工记》中的力学和声学知识，物理通报，1965, (6).

Du Zhengguo, *The mechanical and acoustical knowledge in "Kao Gong Ji"*, Wuli Tongbao, 1965, (6).

范琪，《考工记》的工艺思想，史学月刊，2005, (10).

Fan Qi, *The technological thought of "Kao Gong Ji"*, Shixue Yuekan, 2005, (10).

范志民，《周礼·冬官考工记·画缋》琐谈古代色彩学的萌芽，新美术，1988, (2).

Fan Zhimin, *Short discussion of the germs of the ancient colour doctrine in "Zhou Li·Dong Guan Kao Gong Ji·Hua Hui"*, Xin Meishu, 1988, (2).

方海、景楠，从《考工记》到《鲁班经》: 中国人的设计观 –《建筑与家具》连载 (一)，家具与室内装饰，2011, (7).

Fang Hai, Jing Nan, *From "Kao Gong Ji" to "Lu Ban Jing": Design views of the Chinese*, continuation series 1 of "Jianzhu yu Jiaju" Furniture and in-room decoration, 2011, (7).

方建军，西周磬与《考工记·磬氏》磬制，乐器，1989, (2).

Fang Jianjun, *Stone chimes of Western Zhou dynasty and the manufacture of stone chimes in "Kao Gong Ji·Qing Shi"*, Yueqi, 1989, (2).

方利山，析戴震《考工记图》之 "奇"，中国科技典籍研究 – 第一届中国科技典籍国际会议论文集，郑州: 大象出版社，1998。

Fang Lishan, *Analysis of the "strange" in Dai Zhen's "Kao Gong Ji Tu", Research on the classics of Chinese science and technology*, Proc. of the 1. International conference on the classics of Chinese science and technology, Zhengzhou, Daxiang publ. house, 1998.

傅小龙、王国文，《考工记》"舆人" 篇的设计思想及其启示，科技信息 (科学教研)，2008, (24).

Fu Xiaolong, Wang Guowen, *Design ideas of the chapter "Yu Ren" of "Kao Gong Ji" and its inspiration*, Keji Xinxi (Kexue jiaoyan), 2008, (24).

高怀瑾，中国古代制器质量监督管理 (系列连载六)《考工记》的真实性，上海计量测试，2010, (2).

Gao Huaijin, *Supervision and management of product manufacture quality in ancient China* (continuation series 6) "The authenticity of 'Kao Gong Ji'", Shanghai Jiliang Ceshi, 2010, (2).

高美进，《考工记》中的齐国乐器制造业，管子学刊，2004, (3).

Gao Meijin, *Manufacture of musical instruments in the state Qi in "Kao Gong Ji"*, Guanzi Xuekan, 2004, (3).

关增建，《考工记》角度概念刍议，自然辩证法通讯，2000, (2).

Guan Zengjian, *A humble opinion on the concept of angle in "Kao Gong Ji"*, Ziran Bianzhengfa Tongxun, 2000, (2).

关增建、王胜利，《考工记》里的角度体系，中国科技典籍研究 – 第二届中国科技典籍国际会议论文集，郑州:大象出版社，2003。

Guan Zengjian, Wang Shengli, *The angle system in "Kao Gong Ji", Research on the classics of Chinese science and technology*, Proc. of the 1. International conference on the classics of Chinese science and technology, Zhengzhou, Daxiang publ. house, 1998.

关增建，略谈中国历史上的弓体弹力测试，自然辩证法通讯，1994, (6).

Guan Zengjian, *Short talk on the measurement of the spring force of bows in Chinese history*, Ziran Bianzhengfa Tongxun, 1994, (6).

郭沫若，考工记的年代与国别，《开明书店二十周年纪念文集》，开明书店，1947;《天地玄黄》，大孚出版公司，1947;《沫若文集》第 16 卷，北京:人民文学出版社，1962。

Guo Moruo, *Time and location of origin of Kao Gong Ji*, Literature collection commemorating the 20th anniversary of Kaiming bookshop, Kaiming shudian, 1947, "Tiandi Xuanhuang", Dafu chuban gongsi, 1947; "Collected works of Guo Moruo", vol. 16, Beijing: Renmin wenxue chubanshe, 1962.

郭书春，《管子》与中国古代数学 – 《考工记》时代的中国数学之一，中国科技典籍研究 – 第一届中国科技典籍国际会议论文集，郑州:大象出版社，1998。

Guo Shuchun, *"Guan Zi" and mathematics in ancient China – China's mathematics in the time of "Kao Gong Ji"*, part 1, Research on the classics of Chinese science and technology, Proc. of the 1. International conference on the classics of Chinese science and technology, Zhengzhou, Daxiang publ. house, 1998.

原田淑人，论周官考工记的性质及其写作年代 – 史海片帆 (2)，(日) 圣心女子大学论丛，1967, (30)。

Harada Yoshihito, *On the character of the Zhou Guan Kao Gong Ji and the time of its origin*, Shikai Hempan (2), Collection of publications of Keishin Womens College, 1967, (30).

原田淑人，周官考工记的考古学的检讨，(东京) 东方学报，1936，第 6 册。

Harada Yoshihito, *Archaeological investigations on Zhou Guan Kao Gong Ji*, (Tokyo), Toho Gakuho, 1936, vol. 6.

林巳奈夫，《周礼·考工记》的车制，《中国古代科学技术史研究》，(京都) 东方学报，1959, (30)。

Hayashimi Nao, *Manufacture of chariots in "Zhou Li· Kao Gong Ji", "Research on history of ancient Chinese science and technology"*, Kyoto, Toho Gakuho, 1959, (30).

林巳奈夫，中国古代的祭玉·瑞玉，(京都) 东方学报，1969, (40).

Hayashimi Nao, *The ritual jade Ruiyu of ancient China*, (Kyoto) Toho Gakuho, 1969, (40).

何唐坤，"六齐"之管窥，《科技史文集》第15辑，上海:上海科学技术出版社，1989。

He Tangkun, *A humble investigation of "Six Mixture Ratios"*, Collected works on the history of science and technology, vol. 15, Shanghai: Shanghai kexue jishu chubanshe, 1989.

贺从容，《考工记》模式与希波丹姆斯模式中的方格网之比较，建筑学报，2007, (2).

He Congrong, *Comparison between the grid squares in "Kao Kong Ji" and at Hippotames*, Jianzhu Xuebao, 2007, (2).

贺圣迪，《考工记》与科学思想，上海大学学报 (社会科学版)，1989, (1).

He Shengdi, *The "Kao Gong Ji" and scientific thought*, Shanghai Daxue Xuebao (Shehui kexue ban), 1989, (1).

贺双非，《考工记》的年代、作者与价值，湖南城市学院学报，2009, (4).

He Shuangfei, *The time, the author and the value of "Kao Gong Ji"*, Hunan Chengshi Xueyuan Xuebao, 2009, (4).

侯过，考工记的"轮人"，学术研究，1958, (7).

Hou Guo, *The manufacturer of wheels in "Kao Gong Ji"*, Xueshu Yanjiu, 1958, (7).

后德俊，楚文物与《考工记》的对照研究，中国科技史料，1996, (1).

Hou Dejun, *Comparative research between cultural relics of Chu state and "Kao Gong Ji"*, Zhongguo Keji Shiliao, 1996, (1).

瓠芦，《考工记》磬制的研究，音乐杂志，1928, (6).

Hu Lu, *Investigation of the manufacture of stone chimes in "Kao Gong Ji"*, Yinyue Zazhi, 1928, (6).

华觉明、高宣，"和"的技术观 – 从《考工记》到《天工开物》，中国科技典籍研究 – 第一届中国科技典籍国际会议论文集，郑州:大象出版社，1998。

Hua Jueming, Gao Xuan, *The technical view of the concept "he" – From "Kao Gong Ji" to "Tian Gong Kai Wu", Research on the classics of Chinese science and technology*, Proc. of the 1. International conference on the classics of Chinese science and technology, Zhengzhou, Daxiang publ. house, 1998.

华觉明、贾云福，先秦编钟设计制作的探讨 – 兼论《考工记》"遂"的部位与作用，自然辩证法通讯，1981, (5).

Hua Jueming, Jia Yunfu, *Discussion of the design and manufacture of bell sets in pre-Qin time, Explanation of the location and the function of "sui"*, Ziran Bianzhengfa Tongxun, 1981, (5).

华觉明、王玉柱、朱迎善，商周青铜器合金配制和"六齐"论释，第三届国际中国科学史讨论会论文集，北京:科学出版社，1990。

Hua Jueming, Wang Yuzhu, Zhu Yingshan, *Preparation of the bronze alloy of spokes in Shang and Zhou dynasty and explanation of the "Six Mixture Ratios"*, Proc. of the 3. International conference on the history of Chinese science, Beijing: Kexue chubanshe, 1990.

黄建军、于希贤，《周礼·考工记》与元大都规划，文博，2002, (3).

Huang Jianjun, Yu Xixian, *"Zhou Li· Kao Gong Ji" and the planning of the capital Dadu of Yuan dynasty*, Wenbo, 2002, (3).

Hulsewé, A.F.P. 史四维，*Form and function in the evolution of the wooden wheel* (演变中的木轮的形式和功能)，中国科技史探索 (Research on the history of Chinese science and technology)，上海：上海古籍出版社 (Shanghai: Shanghai guji chubanshe), 1982.

焦泽阳，周"礼制"与《考工记·匠人营国》对早期都城形态的影响，城市规划学刊，2012, (1).

Jiao Zeyang, *Influence of "rites system" of Zhou dynasty and of "Kao Gong Ji·Jiangren Yingguo" on the design of early capitals*, Chengshi Guihua Xuekan, 2012, (1).

金健民，《考工记》成书之渊源与齐国科学技术思想，中国科技典籍研究 – 第一届中国科技典籍国际会议论文集，郑州：大象出版社，1998。

Jin Jianmin, *Origin of "Kao Gong Ji" and the scientific-technological thought in the Qi state*, Research on the classics of Chinese science and technology, Proc. of the 1. International conference on the classics of Chinese science and technology, Zhengzhou, Daxiang publ. house, 1998.

近藤光男，戴震的考工记图 – 科学思想史的考察，东方学，1955.9–1956.1, 第 11 辑。

Kondo Mitsuo, *Dai Zhen's Illustrated Kao Gong Ji, Investigations on the history of scientific thought*, Tohogaku, 1955.9–1956.1, vol. 11.

老亮，《考工记》中的强度知识，力学与实践，1988, (1).

Lao Liang, *The knowledge on strength in "Kao Gong Ji"*, Lixue yu Shijian, 1988, (1).

乐爱国，《考工记》与《管子》的数量观念之比较，中国科技典籍研究 – 第一届中国科技典籍国际会议论文集，郑州：大象出版社，1998。

Le Aiguo, *Comparison of quantitative ideas in "Kao Gong Ji" and "Guan Zi"*, Research on the classics of Chinese science and technology, Proc. of the 1. International conference on the classics of Chinese science and technology, Zhengzhou, Daxiang publ. house, 1998.

李崇州，试探《考工记》中"耒"的形制，农业考古，1995, (3).

Li Chongzhou, *An attempt to discuss the shape of spades "Lei" in "Kao Gong Ji"*, Nongye Kaogu, 1995, (3).

李锋，《考工记》成书西汉时期管窥，郑州大学学报 (哲学社会科学版)，1999, (2).

Li Feng, *A humble view on the origin of "Kao Gong Ji" in Western Han dynasty*, Zhengzhou Daxue Xuebao (Zhexue shehui kexue ban), 1999, (2).

李红云，从《考工记》看齐国乐器制造业，民族艺术，2011, (3).

Li Hongyun, *From "Kao Gong Ji" viewing the manufacture of musical instruments in state Qi*, Minzu Yishu, 2011, (3).

李民、王星光，略论《考工记》车的制造及工艺，河南师范大学学报 (哲学社会科学版)，1985, (2).

Li Min, Wang Xingguang, *Short discussion of the manufacture and technology of chariots in "Kao Gong Ji"*, Henan Shifan Daxue Xuebao (Zhexue shehui kexue ban), 1985, (2).

李民、王星光，略论《考工记》车的制造规范，河南师范大学学报 (哲学社会科学版)，1987, (1).

Li Min, Wang Xingguang, *Short discussion on the standards for the manufacture of chariots in "Kao Gong Ji"*, Henan Shifan Daxue Xuebao (Zhexue shehui kexue ban), 1987, (1).

李强，清儒对《考工记》车制的研究，中国科技典籍研究 – 第一届中国科技典籍国际会议论文集，郑州:大象出版社，1998。

Li Qiang, *Researches of confucianist scholars in Qing dynasty on the manufacture of chariots in "Kao Gong Ji"*, Research on the classics of Chinese science and technology, Proc. of the 1. International conference on the classics of Chinese science and technology, Zhengzhou, Daxiang publ. house, 1998.

李秋芳，20 世纪《考工记》研究综述，中国史研究动态，2004，(5).

Li Qiufang, *Summary of researches on "Kao Gong Ji" in 20. century*, Zhongguo Shi Yanjiu Dongtai, 2004, (5).

李秋芳，林希逸《鬳斋考工记解》及其价值，河南师范大学学报 (哲学社会科学版)，2011, (2).

Li Qiufang, *Lin Xiyi's "Explanation of Kao Gong Ji from the studio of Yan vessels" and its value*, Henan Shifan Daxue Xuebao (Zhexue shehui kexue ban), 2011, (2).

李秋芳，宋代《考工记》研究述论，巢湖学院学报，2010, (2).

Li Qiufang, *Discussion of the research on "Kao Gong Ji" in Song dynasty*, Chaohu Xueyuan Xuebao, 2010, (2).

李双，中国古代造型设计中的功能因素及其理论 –《考工记》阅读手记，装饰，1987, (2).

Li Shuang, *Function factors in the shape design in Chinese antiquity and their theory – Reading notes on "Kao Gong Ji"*, Zhuangshi. 1987, (2).

李宪锋，"人化的自然" – 《考工记》中设计理念的现代性思考，南京艺术学院学报 (美术与设计版)，2008, (3).

Li Xianfeng, *"Humanized nature" – Modern meditation on the design ideas in "Kao Gong Ji"*, Nanjing Yishu Xueyuan Xuebao (Meishu yu sheji ban), 2008, (3).

李学勤，《考工记》与战国兵器铭文的"执齐"，中国科技典籍研究 – 第一届中国科技典籍国际会议论文集，郑州:大象出版社，1998。

Li Xueqin, *"Kao Gong Ji" and "Zhe Ji" inscription on weapons of Warrying States period*, Research on the classics of Chinese science and technology, Proc. of the 1. International conference on the classics of Chinese science and technology, Zhengzhou, Daxiang publ. house, 1998.

李亚明，《考工记》弓矢词语系统考，湖南文理学院学报 (社会科学版)，2007, (2).

Li Yaming, *Exploration of the terminology system for bows and arrows in "Kao Gong Ji"*, Hunan Wenli Xueyuan Xuebao (Shehui kexue ban), 2007, (2).

李亚明，《考工记》名辨，文史杂志，2008, (2).

Li Yaming, *Discussion of terms in "Kao Gong Ji"*, Wenshi Zazhi, 2008, (2).

李亚明，《周礼·考工记》"名-量" 同词形的语义基础，淄博师专学报，2007, (1).

Li Yaming, *The semantic basis of word forms for "ming – liang" in "Zhou Li·Kao Gong Ji"*, Zibo Shizhuan Xuebao, 2007, (1).

李亚明，《周礼·考工记》车舆词语系统 (上)，西华大学学报 (哲学社会科学版)，2007, (4).

Li Yaming, *The terminology system for chariots in "Zhou Li·Kao Gong Ji" (1. p.)*, Xihua Daxue Xuebao (Zhexue shehui kexue ban), 2007, (4).

李亚明，《周礼·考工记》车舆词语系统 (下)，西华大学学报 (哲学社会科学版)，2007, (5).

Li Yaming, *The terminology system for chariots in "Zhou Li·Kao Gong Ji" (2. p.)*, Xihua Daxue Xuebao (Zhexue shehui kexue ban), 2007, (5).

李亚明，《周礼·考工记》度量衡比例关系考，古籍整理研究学刊，2010, (1).

Li Yaming, *Exploration of the proportionality relations of metrology in "Zhou Li·Kao Gong Ji"*, Guji Zhengli Yanjiu Xuekan, 2010, (1).

李亚明，《周礼·考工记》沟洫词语关系，农业考古，2007, (4).

Li Yaming, *Relations of the terminology for channels in "Zhou Li·Kao Gong Ji"*, Nongye Kaogu, 2007, (4).

李亚明，《周礼·考工记》乐钟词语系统，阜阳师范学院学报 (社会科学版)，2007, (6).

Li Yaming, *The system of terminology for music bells in "Zhou Li·Kao Gong Ji"*, Fuyang Shifan Xueyuan Xuebao (Shehui kexue ban), 2007, (6).

李亚明，《周礼·考工记》色彩词语系统，郧阳师范高等专科学校学报，2007, (4).

Li Yaming, *The system of terms for colours in "Zhou Li·Kao Gong Ji"*, Yunyang Shifan Gaodeng Zhuankexiao Xuebao, 2007, (4).

李亚明，《周礼·考工记》时空词语关系，重庆文理学院学报 (社会科学版)，2007, (4).

Li Yaming, *Relations of the terms for time and space in "Zhou Li·Kao Gong Ji"*, Chongqing Wenli Xueyuan Xuebao (Shehui kexue ban), 2007, (4).

李亚明，《周礼•考工记》手工业原材料词语关系，淮海工学院学报 (社会科学版)，2008, (1).

Li Yaming, *Relations of the terms on raw materials for handicrafts in "Zhou Li·Kao Gong Ji"*, Huaihai Gongxueyuan Xuebao (Shehui kexue ban), 2008, (1).

李亚明，《周礼·考工记》营国词语关系，殷都学刊，2007, (3).

Li Yaming, *Relations of the terms for city building*, Yindu Xuekan, 2007, (3).

李亚明，从《周礼·考工记》沟洫关系看我国古代农田水利系统，黄河水利职业技术学院学报，2008, (2).

Li Yaming, *From the relations between different channel types in "Zhou Li·Kao Gong Ji" regarding the system of water policy in ancient China's agriculture*, Huanghe Shuili Zhiye Jishu Xueyuan Xuebao, 2008, (2).

李亚明，论《周礼·考工记》色彩词语系统，通化师范学院学报，2007, (9).

Li Yaming, *On the system of colour terminology in "Zhou Li·Kao Gong Ji"*, Tonghua Shifan Xueyuan Xuebao, 2007, (9).

李亚明，论《周礼·考工记》手工业职官系统的特征，中国石油大学学报（社会科学版），2008, (1).

Li Yaming, *On the characteristics of the official system for artificers in "Zhou Li·Kao Gong Ji"*, Zhongguo Shiyou Daxue Xuebao (Shehui kexue ban), 2008, (1).

李亚明，论《周礼·考工记》玉器词语系统的特征，唐山学院学报，2008, (5).

Li Yaming, *On the characteristics of the terminology system for jade objects in "Zhou Li·Kao Gong Ji"*, Tangshan Xueyuan Xuebao, 2008, (5).

李砚祖，"材美工巧"：《周礼·冬官·考工记》的设计思想，南京艺术学院学报 (美术与设计)，2010, (5).

Li Yanzu, *"Excellent material and perfect technology": Design thought in "Zhou Li·Dong Guan·Kao Gong Ji"*, Nanjing Yishuyuan xuebao (Meishu yu sheji), 2010, (5).

李艳，《考工记》美学研究定位，中国石油大学学报 (社会科学版)，2006, (5).

Li Yan, *Orientation for the aesthetics research on "Kao Gong Ji"*, Zhongguo Shiyou Daxue Xuebao (Shehui kexue ban), 2006, (5).

李玉平，郑玄《周礼注》从历时角度对字际关系的沟通，古汉语研究，2009, (3).

Li Yuping, *The diachronic interpretation of verbal relationships in Zheng Xuan's "Zhou Li Zhu"*, Gu Hanyu Yanjiu, 2009, (3).

李志超，《考工记》与科技训诂，文献，1997, (3); 中国科技典籍研究 – 第一届中国科技典籍国际会议论文集，郑州：大象出版社，1998。

Li Zhichao, *"Kao Gong Ji" and the interpretation of science and technology*, Wenxian, 1997, (3); Research on the classics of Chinese science and technology, Proc. of the 1. International conference on the classics of Chinese science and technology, Zhengzhou, Daxiang publ. house, 1998.

李志超，《考工记》与儒学 – 兼论李约瑟之得失，管子学刊，1996, (4); 中国科技典籍研究 – 第一届中国科技典籍国际会议论文集，郑州：大象出版社，1998。

Li Zhichao, *"Kao Gong Ji" and Confucianism – Discussion of the ideas of J. Needham*, Guanzi Xuekan, 1996, (4); Research on the classics of Chinese science and technology, Proc. of the 1. International conference on the classics of Chinese science and technology, Zhengzhou, Daxiang publ. house, 1998.

李仲达、华觉明、张宏礼，商周青铜器合金成分的考察 – 兼论钟鼎之齐的形成，西北大学学报，1964, (2).

Li Zhongda, Hua Jueming, Zhang Hongli, *Investigation of the alloy composition of bronze objects from Shang and Zhou dynasty – Discussion on the establishing of the compositions for bells and ding vessels*, Xibei Daxue Xuebao, 1964, (2).

梁江，《考工记》新释，美术观察，2002, (2).

Liang Jiang, *New explanation of "Kao Gong Ji"*, Meishu Guancha, 2002, (2).

梁津，周代合金成分考，科学，1925, (10).

Liang Jin, *Investigation of the composition of alloys in Zhou dynasty*, Kexue, 1925, (10).

梁镇海、张业、李和平、魏毅强、孟乃昌，从钟鼎之齐方差分析看《考工记》六齐之成份，太原理工大学学报，1995, 3).

Liang Zhenhai, Zhang Ye, Li Heping, Wei Yiqiang, Meng Naichang, *From a variance analysis of the composition of bells and ding vessels regard the composition of the six mixtures ratios in "Kao Gong Ji"*, Taiyuan Ligong Daxue Xuebao, 1995, (3).

刘斌，《考工记·匠人》和齐国城建，中国科技典籍研究－第一届中国科技典籍国际会议论文集，郑州:大象出版社，1998。

Liu Bin, *"Kao Gong Ji·Jiang Ren" and the building of cities in state Qi*, Research on the classics of Chinese science and technology, Proc. of the 1. International conference on the classics of Chinese science and technology, Zhengzhou, Daxiang publ. house, 1998.

刘春霞、张洪亮，设色之工，画缋之事－散论《考工记》的机械设计美学思想，装饰，2010, (6).

Liu Chunxia, Zhang Hongliang, *Dyeing and the work of painters and embroiderers – Discussion of mechanical designs and aesthetic thoughts in "Kao Gong Ji"*, Zhuangshi, 2010, (6).

刘道广，《考工记》及其研究，装饰，1990, (4).

Liu Daoguang, *"Kao Gong Ji" and related research*, Zhuangshi, 1990, (4).

刘敦愿，《考工记·梓人为笱虡》篇今译及所见雕塑装饰艺术理论，美术研究，1985, (2).

Liu Dunyuan, *Translation of the chapter "The carpenter zi ren manufactures stands for sets of bells in Kao Gong Ji" and the observed art theory of the carved ornaments*, Meishu Yanjiu, 1985, (2).

刘敦愿，《考工记·梓人为笱虡》条所见雕刻装饰理论，山东大学学报 (史)，1962, (2).

Liu Dunyuan, *The observed art theory of the carved ornaments in the chapter "The carpenter zi ren manufactures stands for sets of bells in Kao Gong Ji"*, Shandong Daxue Xuebao (Shi), 1962, (2).

刘广定，再研《考工记》，广西民族学院学报 (自然科学版)，2005, (3).

Liu Guangding, *New investigation of "Kao Gong Ji"*, Guangxi Minzu Xueyuan Xuebao (Ziran kexue ban), 2005, (3).

刘海旺、李京华，三百余件先秦编钟结构制度的统计与分析 – 实物编钟与《考工记》中制度的对比与研究，中国科技典籍研究－第一届中国科技典籍国际会议论文集，郑州:大象出版社，1998。

Liu Haiwang, Li Jinghua, *Statistics and analysis of the structure system of more than 300 bells of bell sets from pre-Qin period – Comparison and investigation of real bell sets with the system in "Kao Gong Ji"*, Research on the classics of Chinese science and technology, Proc. of the 1. International conference on the classics of Chinese science and technology, Zhengzhou, Daxiang publ. house, 1998.

刘洪涛，《考工记》不是齐国官书，自然科学史研究，1984, (4).

Liu Hongtao, *The "Kao Gong Ji" is not an official book of state Qi*, Ziran Kexueshi Yanjiu, 1984, (4).

刘明玉，《考工记》服饰工艺蕴含的“天人合一”思想，艺术与设计(理论)，2009, (9).

Liu Mingyu, *The idea "Heaven and man are one" embodied by the decoration art of "Kao Gong Ji"*, Yishu yu Sheji (Lilun), 2009, (9).

刘明玉，《考工记》服饰染色工艺研究 – 试论“钟氏染羽”，武汉理工大学学报(社会科学版)，2007, (1).

Liu Mingyu, *Research on the technologies of decoration and dyeing in "Kao Gong Ji" – An attempt to discuss the chapter "The official Zhong Shi dyes feathers"*, Wuhan Ligong Daxue Xuebao (Shehui kexue ban), 2007, (1).

刘晓、周明星，《考工记》职业教育史料价值初考，职教论坛，2007, (23).

Liu Xiao, Zhou Mingxing, *Initial investigation of the value of "Kao Gong Ji" as a historic document on the education of professions*, Zhijiao Luntan, 2007, (23).

刘岩，听大师讲析 – 评张道一先生《考工记注译》，美术之友，2005, (4).

Liu Yan, *Hearing the masters lectures and analysis – A critic of Mr. Zhang Dao's "Kao Gong Ji commented and translated"*, Meishu zhi You, 2005, (4).

刘宗超，从艺术学角度看《考工记》，南阳师范学院学报，2005, (1).

Liu Zongchao, *From the point of view of art regard the "Kao Gong Ji"*, Nanyang Shifan Xueyuan Xuebao, 2005, (1).

刘祖慰，《考工记》的哲学背景，中国科技典籍研究 – 第一届中国科技典籍国际会议论文集，郑州:大象出版社，1998。

Liu Zuwei, *The philosophical background of "Kao Gong Ji"*, Research on the classics of Chinese science and technology, Proc. of the 1. International conference on the classics of Chinese science and technology, Zhengzhou, Daxiang publ. house, 1998.

卢静，《考工记》与中国建筑艺术的“正”、“奇”之美，青海社会科学，2006, (5).

Lu Jing, *"Kao Gong Ji" and the aesthetics of "regular" and "strange" in Chinese architectural art*, Qinghai Shehui Kexue, 2006, (5).

路迪民，“六齐”新探，文博，1999, (2).

Lu Dimin, *New examination of the "Six Mixing Ratios" of bronze alloys*, Wenbo, 1999, (2).

罗军，从《考工记》物件词看传统造物的思维方式，南京艺术学院学报 (美术与设计)，2010, (5).

Luo Jun, *From the terms of objects in "Kao Gong Ji" regard the way of thinking of traditional creation*, Nanjing Yishu Xueyuan Xuebao (Meishu yu sheji), 2010, (5).

罗军，论考工记的“和合”思想，大家，2011, (7).

Luo Jun, *On the idea of "harmony" in Kao Gong Ji*, Dajia, 2011, (7).

罗庸，模制考工记车制记，国立历史博物馆丛刊，1926 年 1 卷 1 期；国立中山大学语言历史学研究所周刊，1928 年 4 集 48 期。

Luo Yong, *Note on the manufacture of a model of a chariot according to Kao Gong Ji*, Guoli Lishi Bowuguan Congkan, 1926, vol. 1, issue 1; Guoli Zhongshan Daxue Yuyan Lishixue Yanjiusuo Zhoukan, 1928, 4. collection, issue 48.

罗庸，模制考工记车制述略，考古学论丛，1928, (1).

Luo Yong, *Short report on the manufacture of a model of a chariot according to Kao Gong Ji*, Kaoguxue Luncong, 1928, (1).

水野清，桃氏的青铜剑，考古学杂志，1940, (1–5).

Mizuno Kiyoshi, *The bronze swords of Tao Shi*, Kokogaku Zasshi, 1940, (1–5).

那志良，周礼考工记玉人新注，大陆杂志，1964, (1).

Na Zhiliang, *New comment on the chapter "The jade polisher" in Zhou Li Kao Gong Ji*, Dalu Zazhi, 1964, (1).

大久保庄太郎，周礼考工记，羽衣学园短期大学纪要，1969, (6).

Okubo Shotaro, *Zhou Li Kao Gong Ji*, Notes of the Hagoromo Junior College, 1969, (6).

潘鲁生，《考工记》与《天工开物》的造物思想，美术史论，1990, (4).

Pan Lusheng, *The thought of creation in "Kao Gong Ji" and "Tian Gong Kai Wu"*, Meishu Shilun, 1990, (4).

彭林，《考工记》"数尚六" 现象初探，中国科技典籍研究 – 第三届中国科技典籍国际会议论文集，郑州:大象出版社，2006。

Peng Lin, *First Explorations of the Preference of the Numeral "Six" in "Kao Gong Ji"*, Research on the classics of Chinese science and technology, Proc. of the 3. International conference on the classics of Chinese science and technology, Zhengzhou, Daxiang publ. house, 2006.

彭祖钤，漫话《考工记》，光明日报，1961 年 8 月 5 日 2 版。

Peng Zuling, *Entertaining on "Kao Gong Ji"*, Guangming Ribao, 5.8.1961, 2. edition.

钱慧真，戴震《考工记图注》的名物训诂研究，淮北煤炭师范学院学报 （哲学社会科学版），2008, (2).

Qian Huizhen, *Research on the interpretation of names of objects in Dai Zhen's "Kao Gong Ji Tu Zhu"*, Huaibei Meitan Shifan Xueyuan Xuebao (Zhexue shehui kexue ban), 2008, (2).

磬年，功致为上 – 《考工记》研究笔记，装饰，1990, (4).

Qing Nian, *Function is most important – Research notes on "Kao Gong Ji"*, Zhuangshi, (4).

丘光明，先秦时期质量管理思想措施与法规，考古与文物，1996, (3).

Qiu Guangming, *Ideas, measures and stipulations on the quality management in pre-Qin time*, Kaogu yu Wenwu, 1996, (3).

邱隆，中国最早的度量衡标准器 – 《考工记》·栗氏量，中国计量，2007, (5).

Qiu Long, *The earliest metrological standard of China – "Kao Gong Ji" ·Standard Li Shi Liang*, Zhongguo Jiliang, 2007, (5).

邱隆，中国最早的度量衡标准器 – 《考工记》·栗氏量 (续)，中国计量，2007，(6).

Qiu Long, *The earliest metrological standard of China – "Kao Gong Ji" ·Standard Li Shi Liang (cont.)*, Zhongguo Jiliang, 2007, (6).

Schnell, Wolf H. (维快)，《考工记》和 De Architectura – 两本古籍中的不同技术观点，中国科技典籍研究 – 第一届中国科技典籍国际会议论文集，郑州:大象出版社，1998。

Schnell, Wolf H., *"Kao Gong Ji" and "De Architectura" – Different technical views in two old books, Research on the classics of Chinese science and technology*, Proc. of the 1. International conference on the classics of Chinese science and technology, Zhengzhou, Daxiang publ. house, 1998.

森文，《考工记》工艺美学思想探析，云南艺术学院学报，2011, (2).

Sen Wen, *Examination and analysis of the technological-aesthetical ideas in "Kao Gong Ji"*, Yunnan Yishu Xueyuan Xuebao, 2011, (2).

山东省基建委员会理论小组等，《考工记·匠人》与儒法斗争，建筑学报，1975, (2).

Theory group of the committee of investment construction of Shandong province, *"Kao Gong Ji·chapter Jiang Ren" and the fight between Confucianists and Legalists*, Jianzhu Xuebao, 1975, (2).

商承祚，程瑶田"桃氏为剑"考补正，金陵学报，1938，(1、2).

Shang Chengzuo, Cheng Yaotian, *Examination and correction of the chapter "Tao Shi manufactures swords"*, Jinling Xuebao, 1938, (1,2).

沈长云，谈古官司空之职 – 兼说《考工记》的内容及作成年代，中华文史论丛，1983, (3).

Shen Changyun, *Talking on the tasks of the ancient official Minister of Works – together with a discussion on the content of "Kao Gong Ji" and its time of origin*, Zhonghua Wenshi Luncong, 1983, (3).

沈加锋，《考工记》对中国古代都城中轴线发展过程的影响，新建筑，1989, (2).

Shen Jiafeng, *The influence of "Kao Gong Ji" on the development process of the central axis of ancient Chinese capitals*, Xin Jianzhu, 1989, (2).

沈伊瓦，《考工记》空间认知图景初探 – 以认知心理学、文字语言学为视角，建筑师，2010, (4).

Shen Yiwa, *Initial examination of the cognitive scenery of space in "Kao Gong Ji" – under the point of view of cognitive psychology and semiotic linguistics*, Jianzhushi, 2010, (4).

申永良，《考工记》六齐配方的相图理论研究，冶金丛刊，2000, (2).

Shen Yongliang, *Theoretical investigation of the phase diagrams for the recipes of the Six Mixing Ratios in "Kao Gong Ji"*, Yejin Congkan, 2000, (2).

史宏云，《考工记》与工艺美术的发展，新美术，2011, (3).

Shi Hongyun, *"Kao Gong Ji" and the development of the art of technology*, Xin Meishu, 2011, (3).

史景成，考工记之成书年代考，书目季刊，1971, (3).

Shi Jingcheng, *Investigation of the time of origin of Kao Gong Ji*, Shumu Jikan, 1971, (3).

史念海，《周礼·考工记·匠人营国》的撰著渊源，传统文化与现代化，1998, (3).

Shi Nianhai, *Sources of composing the chapter "Jiang ren Ying Guo" in "Zhou Li· Kao Gong Ji"*, Chuantong Wenhua yu Xiandaihua, 1998, (3).

石平，《考工记》的工艺规范和人体工学的有关原理，浙江工艺美术，1987, (3).

Shi Ping, *Technological standards in "Kao Gong Ji" and related principles of body engineering*, Zhejiang Gongyi Meishu, 1987, (3).

史晓雷，《考工记》中车制问题的两点商榷，广西民族大学学报 (自然科学版)，2008, (4).

Shi Xiaolei, *Two discussion points on the problem of chariot manufacture in "Kao Gong Ji"*, Guangxi Minzu Daxue Xuebao (Ziran kexue ban), 2008, (4).

宋烜，《考工记·匠人》成书年代析，南方文物，1998, (2).

Song Huan, *Analysis of the time of origin of the chapter "Jiang Ren" in "Kao Gong Ji"*, Nanfang Wenwu, 1998, (2).

宋玉立，《考工记·画缋》设计思想释读，济南职业学院学报，2009, (1).

Song Yuli, *Explanation of design ideas in the chapter "Hua Hui" of "Kao Gong Ji"*, Jinan Zhiye Xueyuan Xuebao, 2009, (1).

苏荣誉，《考工记》"六齐" 研究，中国科技典籍研究 – 第一届中国科技典籍国际会议论文集，郑州:大象出版社，1998。

Su Rongyu, *Research on "Six Mixing Ratios" in "Kao Gong Ji"*, Research on the classics of Chinese science and technology, Proc. of the 1. International conference on the classics of Chinese science and technology, Zhengzhou, Daxiang publ. house, 1998.

隋郁，《周礼·考工记·凫氏》两种解读方式之比较，中国音乐，2011, (1).

Sui Yu, *Comparison of two ways to interpret the chapter "Fu Shi" in "Zhou Li· Kao Gong Ji"*, Zhongguo Yinyue, 2011, (1).

孙机，中国古独辀马车的结构，文物，1985, (8).

Sun Ji, *Structure of horse chariots with a pole in ancient China*, Wenwu, 1985, (8).

孙景坛，《周礼》的作者、写作年代及历史意义新探，南京社会科学，1997, (10).

Sun Jingtan, *A new examination of the author, the time of origin and the historical significance of "Zhou Li"*, Nanjing Shehui Kexue, 1997, (10).

孙丽娟、李书谦，《考工记》营国制度与中原地区古代都城布局规划的演变，中原文物，2008, (6).

Sun Lijuan, Li Shuqian, *The system of city building in "Kao Gong Ji" and the evolution of the planning and arrangement of ancient capitals in the region of the Central Plane*, Zhongyuan Wenwu, 2008, (6).

孙庆伟，《考工记•玉人》的考古学研究，考古学研究 (四)，北京:科学出版社，2000。

Sun Qingwei, *Archaeological research on the chapter "Yu ren" of "Kao Gong Ji"*, Kaoguxue Yanjiu (4), Beijing: Kexue chubanshe, 2000.

孙琛，从两周石磬的博谈《考工记》的国别和年代，乐府新声 (沈阳音乐学院学报)，2009, (4).

Sun Zhen, *From two stone chimes of Zhou dynasty comprehensively discuss the origin of location and time of "Kao Gong Ji"*, Yuefu Xinsheng (Shenyang Yinyue Xueyuan Xuebao), 2009, (4).

高田克巳，规矩考 – "周礼考工记" 的考察，大手前女子大学论集，1969, (3).

Takada Katsumi, *Investigation on the right angle, Researches on Zhou Li Kao Gong Ji*, Publications of Otemae Womens College, 1969, p. 3.

高田克巳，规矩考 –"周礼考工记"的考察 (续)，大手前女子大学论集，1969, (4).

Takada Katsumi, *Investigation on the right angle, Researches on Zhou Li Kao Gong Ji (cont.)*, Publications of Otemae Womens College, 1969, p. 4.

高田克巳，规矩考 –"周礼考工记"的考察 (续完)，大手前女子大学论集，1969, (5).

Takada Katsumi, *Investigation on the right angle, Researches on Zhou Li Kao Gong Ji (cont. and end)*, Publications of Otemae Womens College, 1969, p. 5.

覃福勇、崔鹏，景德镇御窑厂祭祀与《考工记》造物原则，美术大观，2011, (2).

Tan Fuyong, Cui Peng, *Sacrifices in the imperial porcelain workshop Jingdezhen and the creation principles of "Kao Gong Ji"*, Meishu Daguan, 2011, (2).

田中淡，先秦山东宫室建筑序说，(京都) 东方学报，1980, (52).

Tanaka Tan, *Introductory explanation of the pre-Qin palace buildings in Shandong*, Kyoto, Toho Gakuho, 1980, (52).

陶正刚，晋都新田与《考工记·匠人》城建的比较，中国科技典籍研究 – 第一届中国科技典籍国际会议论文集，郑州:大象出版社，1998。

Tao Zhenggang, *Comparison between the Jin capital Xintian and the city building in "Kao Gong Ji·Jiang Ren"*, Research on the classics of Chinese science and technology, Proc. of the 1. International conference on the classics of Chinese science and technology, Zhengzhou, Daxiang publ. house, 1998.

田长浒，从现代实验剖析中国古代青铜铸造的科学成就，《科技史文集》第13辑，上海:上海科学技术出版社，1985。

Tian Changhu, *Analysis of the scientific achievements of bronze casting in ancient China by modern experiments*, "Kejishi Wenji", coll. 13, Shanghai: Shanghai kexue jishu chubanshe, 1985.

汪启明，《周礼·考工记》齐语拾补 – 《考工记》为齐人所作再证，古汉语研究，1992, (4).

Wang Qiming, *Collection of words from state Qi in "Zhou Li·Kao Gong Ji" – The "Kao Gong Ji" was repeatedly corrected by people of Qi*, Gu Hanyu Yanjiu, 1992, (4).

汪少华，从《考工记》再看《汉语大字典》的义项漏略，南昌大学学报 (社会科学版)，1996，(4).

Wang Shaohua, *Based on "Kao Gong Ji" regard deficiencies of the dictionary "Hanyu Da Zidian"*, Nanchang Daxue Xuebao (Shehui kexue ban), 1996, (4).

汪少华，"俾輗"考 – 《考工记》名物考证之一，语言研究，2002, (4).

Wang Shaohua, *Investigation on "Bi Ni (pinnacle on the city wall)". Textual critics of object names in "Kao Gong Ji", p. 1*, Yuyan Yanjiu, 2002, (4).

汪少华，从《考工记》看《汉语大字典》的释义失误，传统文化与现代化，1997, (3).

Wang Shaohua, *Based on "Kao Gong Ji" regard erroneous explanations in the dictionary "Hanyu Da Zidian"*, Chuantong Wenhua yu Xiandaihua, 1997, (3).

汪少华，从《考工记》看《汉语大字典》的义项漏略，古汉语研究，1996, (2).

Wang Shaohua, *Based on "Kao Gong Ji" regard deficiencies of the dictionary "Hanyu Da Zidian"*, Gu Hanyu Yanjiu, 1996, (2).

王冬梅，《考工记》工效思想的礼序观考述，作家，2012, (8).

Wang Dongmei, *Examination of the order required by ritual system related to the idea of work efficiency in "Kao Gong Ji"*, Zuojia, 2012, (2).

王冬梅，《考工记》艺术思想的文化生态论，求索，2012, (2).

Wang Dongmei, *The cultural ecology of art thoughts in "Kao Gong Ji"*, Qiusuo, 2012, (2).

王方良，《考工记》设计管理思想探究，电影评介，2006, (16).

Wang Fangliang, *Exploration of the ideas of design management in "Kao Gong Ji"*, Dianying Pingjie, 2006, (16).

王浩军、王坤茜、王芳，《考工记》中四要素在楚国造型艺术中的体现，郑州轻工业学院学报 (社会科学版)，2009, (4).

Wang Haojun, Wang Kunqian, Wang Fang, *Embodiment of the Four Elements in "Kao Gong Ji" in the art of creation in Chu state*, Zhengzhou Qinggongye Xueyuan Xuebao (Shehui kexue ban), 2009, (4).

王琎，中国古代金属原质之化学，科学，1919, (6).

Wang Jin, *Chemistry of metallic raw materials in ancient China*, Kexue, 1919, (6).

王锦光、闻人军，考工记，文史知识，1982, (4).

Wang Jinguang, Wenren Jun, *Kao Gong Ji*, Wenshi Zhishi, 1982, (4).

王玲娟、龙红，《考工记》: 一部很难读便决心把它读通的书 – 关于张道一教授新著《考工记注译》，设计艺术，2005, (4).

Wang Lingjuan, Long Hong, *"Kao Gong Ji" – a book because it is hard to read there is the will to understand it fully – On the new publication of Prof. Zhang Dao "Kao Gong Ji commented and translated"*, Sheji Yishu, 2005, (4).

王梦周，《考工记》玉器设计思想研究 – 试论 "天子圭中必"，武汉理工大学学报 (社会科学版)，2007, (1).

Wang Mengzhou, *Research on the design ideas of jade objects – an attempt to discuss "In the middle of the jade scepter of the son of heaven is wound a silk ribbon"*, Wuhan Ligong Daxue Xuebao (Shehui kexue ban), 2007, (1).

王琦，《考工记》中的和谐设计思想与现代产品设计，科技咨询导报，2007, (16).

Wang Qi, *The idea of harmonious design in "Kao Gong Ji" and the design of modern products*, Keji Zixun Daobao, 2007, (16).

王全太，从《考工记》谈先秦时期的建筑测量，建筑技术，1978, (10).

Wang Quantai, *Based on "Kao Gong Ji" talking about building measurements in pre-Qin times*, Jianzhu Jishu, 1978, (10).

王诗宗，《管子》《考工记》中的数学知识 (科技史话)，人民日报海外版，2002 年 5 月 16 日。

Wang Shizong, *The mathematical knowledge in "Guan Zi" and "Kao Gong Ji" (Talks on the history of science and technology)*, Renmin Ribao Overseas edition, 16.5.2002.

王伟，《考工记》设计思想中的造物与象外之物，艺术探索，2008, (2).

Wang Wei, *The creation and transcendental things in the design ideas of "Kao Gong Ji"*, Yishu Tansuo, 2008, (2).

王伟，从《考工记》中探寻古代设计思想 – 造物与象外之物，设计艺术 (山东工艺美术学院学报)，2009, (1).

Wang Wei, *Search for ancient design ideas in "Kao Gong Ji" – Creation and transcendental things*, Sheji Yishu (Shandong Gongyi Meishu Xueyuan Xuebao), 2009, (1).

王希明，《考工记》的现实教育意义，山东教育学院学报，1990, (2).

Wang Ximing, *The practical educational significance of "Kao Gong Ji"*, Shandong Jiaoyu Xueyuan Xuebao, 1990, (2).

王晓玮、黄瑞，考工记与当代职业教育 –《考工记》中的造型与色彩，中国科教创新导刊，2011, (7).

Wang Xiaowei, Huang Rui, *"Kao Gong Ji" and contemporary professional education – Shaping and colours in "Kao Gong Ji"*, Zhongguo Kejiao Chuangxin Daokan, 2011, (7).

王燮山，《考工记》及其中的力学知识，物理通报，1959, (5).

Wang Xieshan, *"Kao Gong Ji" and the related knowledge on mechanics*, Wuli Tongbao, 1959, (5).

王燮山，《考工记》力学综论，华北电力大学学报 (社会科学版)，1998, (1).

Wang Xieshan, *Comprehensive discussion of mechanics in "Kao Gong Ji"*, Huabei Dianli Daxue Xuebao (Shehui kexue ban), 1998, (1).

王燮山，《考工记》中的声学知识，物理通报，1959, (5).

Wang Xieshan, *The acoustic knowledge in "Kao Gong Ji"*, Wuli Tongbao, 1959, (5).

王燮山，从《考工记》看我国古代的物理学，物理教学，1959, (2).

Wang Xieshan, *Based on "Kao Gong Ji" regarding physics of ancient China*, Wuli Jiaoxue, 1959, (2).

王星光、符奎，徐光启《考工记解》探析，复旦学报 (社会科学版)，2011, (4).

Wang Xingguang, Fu Kui, *Exploration of "Kao Gong Ji explained" by Xu Guangqi*, Fudan Xuebao (Shehui kexue ban), 2011, (4).

王渝生，《考工记》在中国数学史上的地位，中国科技典籍研究 – 第一届中国科技典籍国际会议论文集，郑州:大象出版社，1998。

Wang Yusheng, *The place of "Kao Gong Ji" in the history of Chinese mathematics*, Research on the classics of Chinese science and technology, Proc. of the 1. International conference on the classics of Chinese science and technology, Zhengzhou, Daxiang publ. house, 1998.

王兆祥，《考工记》《营造法式》《工程作法》 – 城市建筑科学技术的发展与进步，中国房地产，2006, (4).

Wang Zhaoxiang, *"Kao Gong Ji", "Ying Zao Fa Shi" and "Gong Cheng Zuo Fa" – Development and progress in science and technology of city building*, Zhongguo Fangdichan, 2006, (4).

王之润，浅析《考工记》对现代设计的影响，才智，2009, (18).

Wang Zhirun, *Short analysis of the influence of "Kao Gong Ji" on modern design*, Caizhi, 2009, (18).
温增源，虚实相应与审美联觉 – 《考工记》中一个杰出的音乐审美观，人民音乐，1989, (5).
Wen Zengyuan, *Correspondence between emptiness and reality and the aesthetical synaesthesia – an outstanding musical aesthetical view in "Kao Gong Ji"*, Renmin Yinyue, 1989, (5).
闻人军，"磬折" 的起源与演变，杭州大学学报 (自然科学版)，1986, (2).
Wenren Jun, *Origin and evolution of "Qing Zhe"*, Hangzhou Daxue Xuebao (Ziran kexue ban), 1986, (2).
闻人军，《考工记》成书年代新考，文史，1984, (23).
Wenren Jun, *New exploration on the time of origin of "Kao Gong Ji"*, Wenshi, 1984, (23).
闻人军，《考工记》齐尺考辨，考古，1983, (1).
Wenren Jun, *Investigation of the measure Chi of state Qi in "Kao Gong Ji"*, Kaogu, 1983, (1).
闻人军，《考工记》磬制倨句考，浙江省历史学会会刊，1981，第1辑。
Wenren Jun, *Investigation of the angle Ju Gou at the manufacture of stone chimes in "Kao Gong Ji"*, Zhejiang Sheng Lishi Xuehui Huikan, 1981, coll. 1.
闻人军，《考工记》中的兵器学，锦州师院学报 (哲学社会科学版)，1987, (2).
Wenren Jun, *Weapon expertise in "Kao Gong Ji"*, Jinzhou Shiyuan Xuebao (Zhexue shehui kexue ban), 1987, (2).
闻人军，《考工记》中的流体力学知识，自然科学史研究，1984, (1).
Wenren Jun, *The knowledge on hydraulics in "Kao Gong Ji"*, Ziran Kexueshi Yanjiu, 1984, (1).
闻人军，《考工记》中声学知识的数理诠释，杭州大学学报 (自然科学版)，1982, (4).
Wenren Jun, *Mathematical interpretation of the acoustic knowledge in "Kao Gong Ji"*, Hangzhou Daxue Xuebao (Ziran kexue ban), 1982, (4).
闻人军，最优化设计 (《考工记》中的六齐·兵车·弓箭与最优化设计)， 中国科技报，1986年4月21日。
Wenren Jun, *Optimum design* (*The Six Mixing Ratios, war chariots, bows and arrows and the optimum design in "Kao Gong Ji"*), Zhongguo Keji Bao, 21.4.1986.
吴承仕，释车 – 三礼名物之一 (上)，国学论衡，1936, (7).
Wu Chengshi, *Explanation of chariots – First of three ritual terms* (*1. part*), Guoxue Lunheng, 1936, (7).
吴点明，试论《考工记》科技思想的人文关怀情结，北京科技大学学报 (社会科学版)，2010, (2).
Wu Dianming, *An attempt to discuss the complex of cultural solicitousness in the scientific-technological ideas of "Kao Gong Ji"*, Beijing Keji Daxue Xuebao (Shehui kexue ban), 2010, (2).
吴来明，"六齐"、商周青铜器化学成分及其演变的研究，文物，1986, (11).

Wu Laiming, *Research on the "Six Mixing Ratios", the chemical composition of bronze objects of Shang and Zhou dynasty and its evolution*, Wenwu, 1986, (11).

武鹏，《考工记》对我国科技发展的意义浅析，南昌教育学院学报，2012, (3).

Wu Peng, *Short analysis of the significance of "Kao Gong Ji" for the development of science and technology in China*, Nanchang Jiaoyu Xueyuan Xuebao, 2012, (3).

武廷海、戴吾三，"匠人营国" 的基本精神与形成背景初探，城市规划，2005, (2).

Wu Tinghai, Dai Wusan, *Initial exploration of the fundamental spirit of the chapter "Jiang Ren Ying Guo" and of the background of its emergence*, Chengshi Guihua, 2005, (2).

吴志坚，浅析春秋时期《考工记》工艺设计教育思想的特征，安徽文学，2012, (5).

Wu Zhijian, *Short analysis of the characteristical educational ideas of technology and design in "Kao Gong Ji" during the Chunqiu period*, Anhui Wenxue, 2012, (5).

肖德武，齐科技文化的意义及其科学发展意蕴 – 从《考工记》、《管子》和《齐民要术》三部名著谈起，齐鲁师范学院学报，2011, (2).

Xiao Dewu, *The significance of science, technology and culture in state Qi and the way of development of science – discussed on the basis of the three works "Kao Gong Ji", "Guan Zi" and "Qi Min Yao Shu"*, Qilu Shifan Xueyuan Xuebao, 2011, (2).

肖屏，《考工记》设计思想探析，武汉科技学院学报，2005, (7).

Xiao Ping, *Exploration and analysis of the design ideas in "Kao Gong Ji"*, Wuhan Keji Xueyuan Xuebao, 2005, (7).

肖屏，《考工记》设计思想研究，湖北美术学院学报，2005, (4).

Xiao Ping, *Research on the design ideas in "Kao Gong Ji"*, Hubei Meishu Xueyuan Xuebao, 2005, (4).

肖世孟，《考工记》设色规范研究，装饰，2009, (5).

Xiao Shimeng, *Research on the standards for choosing colours in "Kao Gong Ji"*, Zhuangshi, 2009, (5).

晓雅，《考工记营国制度研究》简介，城市规划，1988, (6).

Xiao Ya, *Short introduction into "Research on the system of city building in Kao Gong Ji"*, Chengshi Guihua, 1988, (6).

邢文，《考工记》所见色彩思想与相关工艺，中国科技典籍研究 – 第一届中国科技典籍国际会议论文集，郑州:大象出版社，1998。

Xing Wen, *Ideas on colours in "Kao Gong Ji" and related technologies*, Research on the classics of Chinese science and technology, Proc. of the 1. International conference on the classics of Chinese science and technology, Zhengzhou, Daxiang publ. house, 1998.

徐碧珺，从《考工记》看中国传统设计技术观，青年文学家，2009, (9).

Xu Bijun, *From "Kao Gong Ji" regarding the traditional Chinese design and technology*, Qingnian Wenxuejia, 2009, (9).

徐冬树，礼制禁限下的技术知识 – 《考工记》工艺传承的固化趋向，创意与设计，2012, (1).

Xu Dongshu, *Technological knowledge under the restriction of the ritual system – The tendency of congealing in the dissemination of the technologies in "Kao Gong Ji"*, Chuangyi yu Sheji, 2012, (1).

徐勤，《考工记·梓人为简虡》散论梓人•梓庆，上海工艺美术，1994, (4).

Xu Qin, *"Kao Gong Ji Zi Ren Wei Sun Ju", Detailed discussion of the carpenter Zi Ren·Zi Qing*, Shanghai Gongyi Meishu, 1994, (4).

许晓燕，材美工巧 – 《考工记》造物思想的现代启示，艺术教育，2011, (7).

Xu Xiaoyan, *Excellent material, perfect technology – a modern manifestation of the creation ideas in "Kao Gong Ji"*, Yishu Jiaoyu, 2011, (7).

宣兆琦，《考工记》的国别和成书年代，自然科学史研究，1993, (4).

Xuan Zhaoqi, *Location and time of origin of "Kao Gong Ji"*, Ziran Kexueshi Yanjiu, 1993, (4).

薮内清，考工记，《世界大百科事典》，平凡社，1974 年，第 10 册.

Yabuuchi Kiyoshi, *Kao Gong Ji*, "World encyclopaedia", Heibonsha, 1974, vol. 10.

杨恒、章倩励，《考工记》建筑设计理论研究 – 匠人建国、营国的设计思想，美与时代，2007, (9).

Yang Heng, Zhang Qianli, *Research on the theory of building design in "Kao Gong Ji" – Design ideas in the chapter "Jiang ren Jian Guo, Ying Guo"*, Mei yu Shidai, 2007, (9).

杨向奎，周礼内容的分析及其制作时代，山东大学学报，1954, (1).

Yang Xiangkui, *Analysis of the content of "Zhou Li" and the time of its origin*, Shandong Daxue Xuebao, 1954, (1).

杨英杰，先秦战车形制考述，辽宁师范大学学报 (社会科学版)，1984, (2).

Yang Yingjie, *Exploration of the forming of pre-Qin war chariots*, Liaoning Shifan Daxue Xuebao (Shehui kexue ban), 1984, (2).

姚智辉、范云峰，《考工记》"戈体已倨已句二病" 新探，中原文物，2010, (2).

Yao Zhihui, Fan Yunfeng, *A new exploration of the two deficiencies of too large and too small angles between the knives Yuan and Hu on halberds in "Kao Gong Ji"*, Zhongyuan Wenwu, 2010, (2).

尹屾，浅析《考工记》传统设计思想对现代设计的影响，数位时尚 (新视觉艺术)，2009, (1).

Yin Shen, *Short analysis of the influence of traditional design ideas in "Kao Gong Ji" on modern design*, Shuwei Shishang (Xin shijue yishu), 2009, (1).

吉田光邦，对《周礼•考工记》的考察，《中国古代科学技术史研究》，(京都) 东方学报，1959, (30).

Yoshida Mitsukuni, *Investigation of "Zhou Li· Kao Gong Ji"*, "Researches on the history of ancient Chinese science and technology", (Kyoto), Toho gakuho, 1959, (30).

于元元，论《考工记》的编纂特色，档案学通讯，2006, (5).

Yu Yuanyuan, *On the characteristics of compilation of "Kao Gong Ji"*, Dang'anxue Tongxun, 2006, (5).

曾秀芳，郑珍对《考工记》车舆形制的考订，贵州文史丛刊，2011，(2).

Zeng Xiufang, *The textual criticism of Zheng Zhen on the forming of chariots in "Kao Gong Ji"*, Guizhou Wenshi Congkan, 2011, (2).

湛群，《考工记》五行思想与传统工艺美术，安徽农业大学学报 (社会科学版)，2007, (5).

Zhan Qun, *The thought of the Five Phases in "Kao Gong Ji" and the traditional technologies and arts*, Anhui Nongye Daxue Xuebao (Shehui kexue ban), 2007, (5).

张洪亮，天时地气材美工巧 – 散论《考工记》的机械设计美学思想，广东工业大学学报 (社会科学版)，2010, (2).

Zhang Hongliang, *Time of heaven and Qi of earth, excellent material and perfect technology – detailed discussion on the aesthetical ideas of mechanical designs in "Kao Gong Ji"*, Guangdong Gongye Daxue Xuebao (Shehui kexue ban), 2010, (2).

张洪亮，知者创物，巧者和之 – 散论《考工记》的机械设计美学思想，《广东工业大学学报 (社会科学版)》 2010, (6).

Zhang Hongliang, *The wise create things, the clever harmonize them – detailed discussion on the aesthetical ideas of mechanical designs in "Kao Gong Ji"*, Guangdong Gongye Daxue Xuebao (Shehui kexue ban), 2010, (6).

张洪亮、刘春霞，审曲面埶，虚实相生 – 散论《考工记》的机械设计美学思想 (二)，艺术教育，2010, (8).

Zhang Hongliang, Liu Chunxia, *Examine round and even surfaces, emptiness and reality generate each other – detailed discussion on the aesthetical ideas of mechanical designs in "Kao Gong Ji"* (2), Yishu Jiaoyu, 2010, (8).

章鸿钊，中国用锌的起源，科学，1923, (3).

Zhang Hongzhao, *Origin of the use of zinc in China*, Kexue, 1923, (3).

张辉、戴端，《考工记》中所体现的传统造物之美，艺术与设计 (理论)，2009, (3).

Zhang Hui, Dai Duan, *The beauty of traditional creation as embodied in "Kao Gong Ji"*, Yishu yu Sheji (Lilun), 2009, (3).

张健、陈真，《考工记》用漆状况刍议，装饰，2004, (4).

Zhang Jian, Chen Zhen, *A humble opinion on the conditions of using lacquer in "Kao Gong Ji"*, Zhuangshi, 2004, (4).

张乾元，《周礼·考工记·画缋》释读，宿州教育学院学报，2002, (3).

Zhang Qianyuan, *Explanation of the chapter "Hua Hui" in "Zhou Li·Kao Gong Ji"*, Suzhou Jiaoyu Xueyuan Xuebao, 2002, (3).

张蓉，《考工记》营国制度新解 – 与规划模数相关的内容，建筑师，2008, (5).

Zhang Rong, *New explication of the order of city building in "Kao Gong Ji" – content related to planning modules*, Jianzhushi, 2008, (5).

张言梦，略论戴震《考工记图》的编绘及其影响，新美术，2006, (6).

Zhang Yanmeng, *Short discussion on the compiled illustrations in Dai Zhen's "Kao Gong Ji Tu" and their influence*, Xin Meishu, 2006, (6).

张颖，对于"六齐"成份诸见解的思考，阜阳师范学院学报（自然科学版），1994, (1).

Zhang Ying, *Consideration of the opinions on the composition of the "Six Mixing Ratios"*, Fuyang Shifan Xueyuan Xuebao (Ziran kexue ban), 1994, (1).

张越，《考工记》的工艺美学思想，山东社会科学，2005, (6).

Zhang Yue, *The ideas on the aesthetics of technologies in "Kao Gong Ji"*, Shandong Shehui Kexue, 2005, (6).

张鸷中、戴吾三，周代车的社会功能与性质，中国科技典籍研究 – 第一届中国科技典籍国际会议论文集，郑州:大象出版社，1998.

Zhang Zhizhong, Dai Wusan, *The social function and properties of chariots in Zhou period*, Research on the classics of Chinese science and technology, Proc. of the 1. International conference on the classics of Chinese science and technology, Zhengzhou, Daxiang publ. house, 1998.

张子高，六齐别解，清华大学学报（自然科学版），1958, (2).

Zhang Zigao, *A critical interpretation of the six mixing ratios*, Qinghua Daxue Xuebao (Ziran kexue ban), 1958, (2).

赵梓汝，对《考工记》中《匠人》的研究，中国商界（下半月），2010, (11).

Zhao Ziru, *Research on the "Jiang ren (construction worker)" in "Kao Gong Ji"*, Zhongguo Shangjie (2. edition of the month), 2010, (11).

钟正基，《考工记》车的象征意义解析，武汉理工大学学报（社会科学版），2007, (1).

Zhong Zhengji, *Analysis of the symbolic significance of chariots in "Kao Gong Ji"*, Wuhan Ligong Daxue Xuebao (Shehui kexue ban), 2007, (1).

周长山，汉长安城与《考工记》，文物春秋，2001，(4).

Zhou Changshan, *The city of Chang'an of Han dynasty and the "Kao Gong Ji"*, Wenwu Chunqiu, 2001, (4).

周鹏，从出土文物看《考工记》中"准之"内涵，魅力中国，2009, (36).

Zhou Peng, *From excavated culture objects regarding the content of "Zhun Zhi (determination of the volume of the cast material by dipping into water and weighing the replaced water)"*, Meili Zhongguo, 2009, (36).

周始民，《考工记》六齐成分的研究，化学通报，1978, (3).

Zhou Shimin, *Research on the six mixing ratios of components in "Kao Gong Ji"*, Huaxue Tongbao, 1978, (3).

周书灿，《考工记》的科技史料价值新探，殷都学刊，2005, (2).

Zhou Shucan, *New exploration of the value of "Kao Gong Ji" as document on the history of science and technology*, Yindu Xuekan, 2005, (2).

朱建武，浅论《考工记》的美术思想，凯里学院学报，2011, (2).

Zhu Jianwu, *Short discussion of the art thought in "Kao Gong Ji"*, Kaili Xueyuan Xuebao, 2011, (2).

朱志荣、田军，论《考工记》的美学思想，西北大学学报（哲学社会科学版），2009, (5).

Zhu Zhirong, Tian Jun, *On the aesthetical ideas of "Kao Gong Ji"*, Xibei Daxue Xuebao (Zhexue shehui kexue ban), 2009, (5).

邹依仁，《周礼·考工记》中的质量管理，上海社会科学院学术季刊，1985, (2).
Zou Yiren, *Quality management in "Zhou Li·Kao Gong Ji"*, Shanghai Shehui Kexueyuan Xueshu Jikan, 1985, (2).

Editions of "Kao Gong Ji" and Related Research Works

(汉) 郑玄注, (唐) 贾公彦疏，彭林整理，周礼注疏，济南：山东画报出版社，2004.
(Han) Zheng Xuan (commented), (Tang) Jia Gongyan (interpreted), Peng Lin (arranged), *Zhou Li with comment*, Jinan: Shandong huabao chubanshe, 2004.
(汉) 郑玄注，(明) 陈凤梧编，周礼六卷附考工记一卷.
(Han) Zheng Xuan (commented), (Ming) Chen Fengwu (compiled), *Zhou Li in six volumes, added by one volume of Kao Gong Ji.*
(唐) 杜牧注，(民国) 宋联奎校、邵力子署，关中丛书·考工记，陕西通志馆印，1934。
(Tang) Du Mu (commented), (Republic) Song Liankui (corrected), Shao Lizi (arranged), *Guan Zhong Cong Shu·Kao Gong Ji*, Shanxi Tongzhi Guanyin, 1934.
(宋) 林希逸撰，鬳斋考工记解二卷附释音二卷。
(Song) Lin Xiyi (composed), *Explanation of Kao Gong Ji from the studio of Yan vessels*, 2 vol., added by 2 vol. of phonetic explanations.
(宋) 王安石著，周官新义附考工记解，北京：商务印书馆，1935; 中华书局，1985.
(Song) Wang Anshi (composed), *New meaning of Zhou Guan added by Explanation of Kao Gong Ji*, Beijing: Shangwu yinshuguan, 1935, Zhonghua shuju, 1985.
(宋) 张大亨，批点考工记，北京：中华书局，1985.
(Song) Zhang Daheng, *Commented Kao Gong Ji*, Beijing: Zhonghua shuju, 1985.
(元) 吴澄考注；(明) 周梦旸评，批点考工记，北京：中华书局，1991.
(Yuan) Wu Cheng (commented); (Ming) Zhou Mengyang (evaluated), *Commented Kao Gong Ji*, Beijing: Zhonghua shuju, 1991.
(明) 陈兴郊辑，考工记辑注二卷.
(Ming) Chen Xingjiao (compiled), *Kao Gong Ji compiled and commented*, 2 vol.
(明) 程明哲撰，考工记纂注二卷，济南：齐鲁书社，1997.
(Ming) Cheng Mingzhe (composed), *Kao Gong Ji edited and commented*, 2 vol., Jinan, Qilu shushe, 1997.
(明) 郭良翰撰，周礼古本订注五卷、考工记一卷，济南：齐鲁书社，1997.
(Ming) Guo Lianghan (composed), *Old book Zhou Li corrected and commented*, 5 vol., *Kao Gong Ji*, 1 vol., Jinan: Qilu shushe, 1997.
(明) 郭正域批点，考工记二卷，济南：齐鲁书社，1997.
(Ming) Guo Zhengyu (commented), *Kao Gong Ji*, 2 vol., Jinan: Qilu shushe, 1997.
(明) 郎兆玉撰，注释古周礼五卷、考工记一卷，济南：齐鲁书社，1997.

(Ming) Lang Zhaoyu (composed), *Commented and explained old Zhou Li*, 5 vol., *Kao Gong Ji*, 1 vol., Jinan: Qilu shushe, 1997.
(明) 林兆珂撰, 考工记述注 (一)、(二).
(Ming) Lin Zhaoke (composed), *Kao Gong Ji commented*, p. I, II.
(明) 徐光启撰，考工记解二卷，《徐光启著译集》，上海：上海古籍出版社，1983.
(Ming) Xu Guangqi (composed), *Kao Gong Ji explained*, 2 vol., in: *"Works of Xu Guangqi translated and collected"*, Shanghai: Shanghai guji chubanshe, 1983.
(明) 徐昭庆撰，考工记通二卷，济南：齐鲁书社，1997.
(Ming) Xu Zhaoqing (composed), *Complete Kao Gong Ji*, 2 vol., Jinan: Qilu shushe, 1997.
(明) 朱大启撰，考工记辑，不分卷.
(Ming) Zhu Daqi (composed), *Kao Gong Ji collected*, not divided into several volumes.
(清) 陈宗起撰，考工记鸟兽虫鱼释，上海：上海古籍出版社，1996.
(Qing) Chen Zongqi (composed), *Explanation of the fauna in Kao Gong Ji*, Shanghai: Shanghai guji chubanshe, 1996.
(清) 陈衍，考工记辨正三卷、考工记补疏一卷，石遗室丛书.
(Qing) Chen Yan, *Kao Gong Ji corrected*, 3 vol., *Kao Kong Ji Bu Shu*, 1 vol., Shi Yi Shi Cong Shu.
(清) 程瑶田，皇清经解，学海堂本.
(Qing) Cheng Yaotian, *Explained classics from imperial Qing dynasty*, Xue Hai Tang Ben.
(清) 戴震著，程瑶田批注，考工记图二卷，北京：商务印书馆，1935, 1955.
(Qing) Dai Zhen (composed), Chen Yaotian (commented), *Illustrated Kao Gong Ji*, 2 vol., Beijing: Shangwu yinshuguan, 1935, 1955.
(清) 东山主人，新辑各国政治艺学全书·制造·考工记要：卷一至卷十七。
(Qing) Dongshan Zhuren, *New collected world encyclopaedia of politics, art and science·Manufacture·Essence of Kao Gong Ji*, Vol. 1 to 17.
(清) 方苞撰，周官析疑·考工记析疑，济南：齐鲁书社，1997; 上海：上海古籍出版社，1996。
(Qing) Fang Bao (composed), *Dispel of doubts in Zhou Guan·Dispel of doubts in Kao Gong Ji*, Jinan: Qilu shushe, 1997; Shanghai: Shanghai guji chubanshe, 1996.
(清) 江永，周礼疑义举要，北京：中华书局，1985.
(Qing) Jiang Yong, *Discussion of problems in Zhou Li*, Beijing: Zhonghua shuju, 1985.
(清) 吕调阳撰，考工记考，上海：上海古籍出版社，1996.
(Qing) Lü Diaoyang (composed), *Investigation of Kao Gong Ji*, Shanghai: Shanghai guji chubanshe, 1996.
(清) 阮元撰，考工记车制图解，上海：上海古籍出版社，1996.
(Qing) Yuan Yuan (composed), *Illustrated explanation of the manufacture of chariots in Kao Gong Ji*, Shanghai, Shanghai guji chubanshe, 1996.
(清) 孙诒让，周礼正义，北京：中华书局，1987.
(Qing) Sun Yirang, *Correct meaning of Zhou Li*, Beijing: Zhonghua shuju, 1987.
(清) 王宗涑撰，孙诒让校，考工记考辨，上海：上海古籍出版社，1996.

(Qing) Wang Zongsu (composed), Sun Yirang (corrected), *Investigation of Kao Gong Ji*, Shanghai: Shanghai guji chubanshe, 1996.

(清) 袁俊德，富强斋丛书续全集·工政·考工记要：卷一至卷十七。

(Qing) Yuan Junde, *Complete works from Fu Qiang studio, Administration of artificers·Essence of Kao Gong Ji*, vol. 1 to 17.

(清) “钦定” 四库全书荟要·礼经会元考工记解，长春：吉林出版集团有限责任公司，2005.

(Qing) *Encyclopaedia Si Ku Quan Shu, made on imperial order·Explanation of Kao Gong Ji*, Changchun: Jilin chuban jituan youxian zeren gongsi, 2005.

(英国) 玛体生著，考工记要 (卷一至卷十七).

(England) Matthiessen (composed), *Essence of Kao Gong Ji*, vol. 1 to 17.

戴吾三，考工记图说，济南：山东画报出版社，2003.

Dai Wusan, *Kao Gong Ji Tu Shuo (Illustrated and explained Kao Gong Ji)*, Jinan, Shandong huabao chubanshe, 2003.

郭超，夏于全主编，传世名著百部·考工记、新仪象法要、数书九章，北京：蓝天出版社，1999。

Guo Chao, Xia Yuquan (main ed.), *Hundred famous works of passed on masters ·Kao Gong Ji, Xin Yi Xiang Fa Yao·Shu Shu Jiu Zhang*, Beijing: Lantian chubanshe, 1999.

贺业钜，考工记营国制度研究，北京：中国建筑工业出版社，1985.

He Yeju, *Research on the system of city building in Kao Gong Ji*, Beijing: Zhongguo jianzhu gongye chubanshe, 1985.

林尹，周礼今注今译，北京：商务印书馆，1972; 书目文献出版社，1985。

Lin Yin, *Zhou Li with modern comment and translation*, Beijing: Shangwu yinshuguan, 1972; Shumu wenxian chubanshe, 1985.

刘道广等著，图证考工记，南京：东南大学出版社，2012.

Liu Daoguang et al., *Illustrated and corrected Kao Gong Ji*, Nanjing: Dongnan Daxue Chubanshe, 2012.

王云五主编，戴震著，万有文库，考工记图，北京：商务印书馆，1935.

Wang Yunwu (main ed.), Dai Zhen (composed), *Wan you Wen Ku, Illustrated Kao Gong Ji*, Beijing: Shangwu yinshuguan, 1935.

Wenren Jun, *Ancient Chinese Encyclopaedia of Technology: Translation and Annotation of Kao Gong Ji (The Artificer's Record)*, London and New York: Routledge, 2013.

闻人军，国学经典导读·考工记，北京：中国国际广播出版社，2008, 2011。

Wenren Jun, *Introduction to the Classics ·Kao Gong Ji*, Beijing: Zhongguo guoji guangbo chubanshe, 2008, 2011.

闻人军，考工记导读，成都：巴蜀书社，1996.

Wenren Jun, *Introduction to Kao Gong Ji*, Chengdu: Bashu shushe, 1996.

闻人军，考工记导读图译，台北：明文书局，1990.

Wenren Jun, *Illustrated and translated Introduction to Kao Gong Ji*, Taibei: Mingwen shuju, 1990.

闻人军，考工记译注，上海:上海古籍出版社，1993, 2008.

Wenren Jun, *Translated and commented Kao Gong Ji*, Shanghai: Shanghai guji chubanshe, 1993, 2008.

夏纬英，《周礼》书中有关农业条文的解释，北京:农业出版社，1979.

Xia Weiying, *Explanations on the item agriculture in the book "Zhou Li"*, Beijing: Nongye chubanshe, 1979.

张道一，考工记注译，西安:陕西人民美术出版社，2004.

Zhang Daoyi, *Commented and translated Kao Gong Ji*, Xi'an: Shanxi renmin meishu chubanshe, 2004.

周成林，考工记，北京:海豚出版社，2012.

Zhou Chenglin, *Kao Gong Ji*, Beijing: Haitun chubanshe, 2012.

Master Theses

段大龙，《考工记》的"材美""工巧"设计思想及其现实意义，东北师范大学，2007.

Duan Dalong, *The design idea of "excellent material" and "perfect technology" in "Kao Gong Ji" and its practical significance*, Dongbei Shifan Daxue, 2007.

李艳，《考工记》工艺美学思想研究，郑州大学，2005.

Li Yan, *Research on the ideas of technology and aesthetics in "Kao Gong Ji"*, Zhengzhou Daxue, 2005.

林卓萍，《考工记》弓矢名物考，杭州师范学院，2006.

Lin Zhuoping, *Investigation on the designations for bows and arrows*, Hangzhou Shifan Xueyuan, 2006.

刘明玉，《考工记》服饰工艺理论研究，武汉理工大学，2007.

Liu Mingyu, *Research on the theory of decoration technologies in "Kao Gong Ji"*, Wuhan Ligong Daxue, 2007.

马华，"尚质尚用" – 《墨经》和《考工记》中的功能主义思想研究，昆明理工大学，2011.

Ma Hua, *"Quality and function" – Research on the idea of funtionalism in "Mo Jing" and "Kao Gong Ji"*, Kunming Ligong Daxue, 2011.

马樱滨，从理念到实践:论元大都的城市规划与《周礼·考工记》之间的关联，复旦大学，2008。

Ma Yingbin. *From idea to practice: On the plan of the city Dadu in Yuan dynasty and the connection with "Zhou Li·Kao Gong Ji"*, Fudan Daxue, 2008.

孙琛，《考工记·磬氏》验证，中国艺术研究院，2007.

Sun Chen, *Examination of the chapter "Qing Shi" in "Kao Gong Ji"*, Zhongguo Yishu Yanjiuyuan, 2007.

孙洪伟，《考工记》设计思想研究，武汉理工大学，2008.

Sun Hongwei, *Research on the design ideas in "Kao Gong Ji"*, Wuhan Ligong Daxue, 2008.
唐中海，《考工记·玉人》名物训诂与孙疏补，杭州师范学院，2005.
Tang Zhonghai, *Interpretation of the designations of the chapter "Yu Ren" in "Kao Gong Ji" and the commentary and supplements by Sun* (*Yirang*), Hangzhou Shifan Xueyuan, 2005.
王冬梅，对《考工记》工效学意涵的文化解读，苏州大学，2006.
Wang Dongmei, *Explanation of the culture contained in the doctrine of work efficiency in "Kao Gong Ji"*, Suzhou Daxue, 2006.
王梦周，《考工记》玉器设计思想研究，武汉理工大学，2007.
Wang Mengzhou, *Research on the design ideas of jade objects in "Kao Gong Ji"*, Wuhan Ligong Daxue, 2007.
钟正基，《考工记》车的设计思想研究，武汉理工大学，2007.
Zhong Zhengji, *Research on the design ideas for chariots in "Kao Gong Ji"*, Wuhan Ligong Daxue, 2007.

Doctoral Dissertations

戴吾三，齐国科技史研究，中国科学技术大学，1996.
Dai Wusan, Research on the history of science and technology in Qi state, University of Science and Technology of China, 1996.
顾莉丹，《考工记》兵器疏证，复旦大学，2011.
Gu Lidan, *Commentary on and correction of weapons in "Kao Gong Ji"*, Fudan Daxue, 2011.
张言梦，汉至清代《考工记》研究和注释史述论稿，南京师范大学，2005.
Zhang Yanmeng, *Revue on the research on the "Kao Gong Ji" and the history of its explanantion from Han until Qing dynasty*, Nanjing Shifan Daxue, 2005.

Index of Personal and Geographical Names

Index of Subjects

www.ingramcontent.com/pod-product-compliance
Lightning Source LLC
Chambersburg PA
CBHW071323230926
51849CB00002B/298
9789004416901